Three Bags of LADIES CLOTHES & a SNIPER

Translator

Liz Waters

Three Bags of LADIES CLOTHES & a SNIPER
TO THE UKRAINIAN BORDER AND BEYOND

by
JAAP SCHOLTEN

Helena History Press

Copyright 2023 © Jaap Scholten
All rights reserved

Published in the United States by:
HELENA HISTORY PRESS LLC
A division of KKL Publications LLC, Reno, NV USA

H P

www.helenahistorypress.com
Publishing scholarship about and from Central and East Europe

The publisher gratefully acknowledges the support
of the Dutch Foundation for Literature

Nederlands letterenfonds
dutch foundation
for literature

ISBN 978-1-943596-36-2 / Paperback

Copy Editor: Krisztina Kós

Graphic design: Sebastian Stachowski

All photographs in this book are by Jaap Scholten, with the exception of those from the Ukrainian Telegram chat groups and Neal. Pictures of artwork in this book and on the cover were taken by the author in the streets of Ukraine. The author and publisher have done the utmost to reach possible copyright holders. The editing of the photographs is by Martien Frijns and Jaap Scholten.

The map on page 9 is by Ton Markus, Kartomedia.

The quotation on page 56 is from Isaac Babel, *Red Cavalry*, translated by Peter Constantine (WW Norton, NY, 2003).

The quotation on page 257 is from an article by Joseph Roth published in the *Frankfurter Zeitung* on 1 December 1926.

Contents

Preface 1

Part I TO THE UKRAINIAN BORDER 5

Hungarian Uprising 7 The border 9 Sanctions 12 Roma from Transcarpathia 14 Old treaties 17 Sergey Naryshkin 20 Child's shoe 23 The border crossing at Lónya 26 Gelsomina 30 'When the bombs came, I took my babies' 32 Bastard Putin 35 Adopt a roadblock 41 *Russkiy voenny korabl idi na huy!* 43 Mamas in Europe 46 A cat called Tasya 50 Where to find a father like that 33 Killing of the dream 57 The story of a chicken 60 The dramaturgy of a mass murderer 65 As plentiful as Russians 69 Medicine for the people 73 How to make Molotov cocktails 76 Hair dryer 81 List 85 Day of Judgement 87 Lavender and shit 90 Between East and West 94 Exodus 99 A glorious death 104 Father 109 Nuns and kebab 111 Derogatory terms 117 The pain 118

Part II AND BEYOND... 121

God called me 123   Empty-nest syndrome in a time of war 128
Modern Spartans – the stories of survivors in Bucha 133
Karma is a cruel thing 138   On the killing of generals 141
The Cossack mentality 147   The Russian method 152
Putin is fantastic! 159   Staring at the sky 165   Destination unknown 170   Dense forests 174   Hi baby, love you 178   Asking the way 180   Black gold 185   War flag 188   Don't worry, they only bomb the Nazis 191   We sometimes forgot the war was so close 194   Lifesaving microwave 198   How to dress for a war 201   First funeral of the day 205   Honorary Guards 207   Second funeral of the day 213   Tango at Taras Shevchenko 216

Part III THE SNIPER 223

Can you do that from further away? 225 Kyiv cake 230
A christening 235 We're approaching the warzone 240
Chernihiv 244 Make sure they don't take you alive 250
In Russia sin is as boring as virtue is among us 254
The Pyramid Mind 259 On fear 266 A boy called Samuel 272
The baby from Mariupol 277 Three bags of ladies clothes 283

Epilogue 289

Mother Ukraine 291

Sláva Ukrayíni!
Heróyam sláva!

Glory to Ukraine!
To the heroes, glory!

Preface

On day two of the war in Ukraine, 25 February 2022, my niece rang me in the early hours to ask whether I could help her brother-in-law's family get out of the country. They had set off with their three young children at six in the morning from Kyiv, heading for Hungary, navigating their way westwards between the bombardments. I got into the car and drove from our village in southern Hungary to the border with Ukraine to help the young family. In the four days I spent at the Hungary-Ukraine border, I saw the despair in the eyes and stooped shoulders of the refugees, and for the first time in my life I became an activist. The indomitability of the Ukrainian people was contagious, making it impossible for me just to sit on my backside, watching Ukrainians die. I instinctively became involved in the war.

I've lived in Hungary for twenty years, with a Hungarian wife whose father fought the Russians in the streets of Budapest in 1956. After the Russians crushed the revolution he fled first to Germany and then to the Netherlands. In March 2022 I crossed the border into Ukraine for the first time, in the middle of the night, and met Neal, a former marksman with the US Marines. Having served in Iraq and Afghanistan, he had now left his wife, his three children and his landscaping business behind in Alabama to help in Ukraine. For him the bombing of the maternity hospital in Mariupol was the final straw. He walked into the country with two heavy rucksacks. We stayed in touch throughout his time there.

Several months later I travelled to Chernihiv to deliver supplies and to pick up Neal in Kyiv. He'd promised his wife that he wouldn't spend more than three months at war. He'd been training troops at the front – men who only a few months earlier had been lawyers, shopkeepers or computer programmers – as well as carrying out sniping missions behind Russian lines. But his most spectacular and cool-headed act was the evacuation of an American citizen who was stranded in the warzone: a six-day-old baby in an underground hospital in Mariupol.

We were a writer and a sniper, travelling together through a country at war. We spent many hours in my Toyota pickup. Via Bucha, Kozelets, Chernihiv, Kyiv, Lviv, Medyka, Poland and Slovakia, I travelled with Neal back to Budapest. During that road trip he told me frankly and in detail about life in the trenches and about his work; 'forward operating', he called it. Because of those sniping assignments in Russian-occupied territory I've changed his name in this book. As for all the other people mentioned, I deliberately use only their first names, to make them harder to trace. Since the discovery of the massacres in Bucha and Izyum, we all know how ruthlessly the Russians behave.

An old lady I interviewed years ago for my book *Comrade Baron*, a woman condemned to spend most of her adult life in a windowless cellar in Târgu Mureș during communism under Ceaușescu, who was spied upon, shadowed and phone tapped by the Romanian secret police, impressed upon me, 'Jaap, you mustn't write only about the big, dramatic things. It's important to write about the little things, everyday things, if you want people to understand life under a regime like that.' Once again, I've taken her words to heart.

For that book, *Comrade Baron*, I interviewed some fifty victims of communism in Hungary and Romania, many of whom had been persecuted and tortured. As a Western European (I was born and raised close to the German border in the peaceful eastern Netherlands, in the province of Twente, a distinctive region with a strong sense of autonomy), those conversations brought home to me the evil of the Soviet Union and the brutal and refined terror of an autocracy. Immediately after the invasion of 24 February 2022, it was clear to me that the Kremlin was again using all the tried and tested methods of the Soviet Union on a large scale: distortion, propaganda, lies, disappearances, rape, torture, murder.

In the first week after the invasion, this led me to conclude that the most humanitarian response was to help the Ukrainian army. Because only an army can stop the torture, murder and rape by Russian soldiers. Since March 2022 we (a Dutch farmer in Ukraine, a former Dutch banker, a Dutchman who had recently graduated and myself) have been collecting money and sending whatever is needed by direct channels to the front: trauma packs, helmets, night sights, drones, food, tablets, tents, tools and pickup trucks. Sometimes we deliver the pickups in convoy ourselves to the indomitable Ukrainian people. That work has meant I've travelled a great deal over the past year.

In a gallery in Kyiv I asked a professor at the Kyiv School of Economics about the task of writers and artists in times like these. 'Documenting, documenting, documenting,' was his reply. That's what I've done. What began as a logbook that I kept as I travelled, and published on the website of my literary manager Oscar van Gelderen, grew to become a more comprehensive account. Sometimes I had hours to tide over, and therefore time in abundance; at other points I was continually on the move or struggled in adverse circumstances to take notes or to translate the stories of the people involved. Those stories eventually became the basis for passages in this book.

This account is about what I saw around me, about a trip through a beleaguered country, about people I met because of the war, in Ukraine, Poland and Hungary – brave women and uprooted families, writers and violinists, soldiers and an American sniper – about the things they told me and the banalities of a war around the corner from where I live. But above all it's about the astonishing courage and solidarity of the Ukrainians, an example and inspiration to us all.

Jaap Scholten
Budapest, 21 June 2023
www.protectukraine.nl

Part I
To the Ukrainian border

Somogy, Friday morning, 25 February 2022, 11.00 hours

Hungarian Uprising

In the early hours of the morning I receive a call from my niece Madelien, the oldest daughter of my brother Frans. She's five months pregnant and living in New Zealand with a Ukrainian architect. His brother and family left Kyiv at six this morning. Can I help them? They want to get out of the country but don't know where to go. With three small children and a grandmother, they are now attempting to drive to the Hungarian border. Their plans are unclear. Kyiv is being bombed. The whole country has come under fire.

I'm in our village in Hungary, in the province of Somogy, and soon – after getting hold of a Hungarian translation of a letter of invitation for the Ukrainian family – I'll get

into the car and drive to the Ukrainian border, in search of the family from Kyiv. I don't know which border crossing they'll make for: Hungary, Slovakia, Poland or Romania.

I'm horrified by what's happening in Ukraine, a neighbouring country being occupied by brute force. We haven't seen a deployment of armour like this in Europe since the Soviet Russians blasted their way through Budapest with more than a thousand tanks in 1956 (and were ordered to shoot at every house, at every window and at anyone who threw a stone or fired a shotgun).

All this reminds me of that Hungarian Uprising of 1956. I admire the courage of the Ukrainians and feel deeply ashamed that the West is worrying about its own finances while people in Ukraine are fighting and dying.

Záhony, Friday evening, 25 February 2022, 23.50 hours

The border

After calling at Budapest I travel east to Beregsurány, the nearest crossing on the border with Ukraine. On the motorway I get stuck in a traffic jam, behind a big military convoy of huge articulated trucks that are travelling in the same direction. The north-east of Hungary, which borders on Slovakia and Ukraine, is the most poverty-stricken part of the country; as soon as you leave the main highway you're struck by the depth of the surrounding darkness, by the endless stretches of nothingness with not a single light, lit-

tle traffic and high fences around the gardens in the villages. You watch your surroundings get poorer. Even the trees alongside the road are mere driftwood.

Throughout the trip, which takes about seven hours, I listen to the radio, mainly the BBC, since the vast majority of people in Ukraine receive it – a tribute to British tenacity. A Dutch friend in Kyiv, Dirck, sends me a message saying he's sheltering in his cellar with his wife and child, along with supplies of food and water. The phone makes my car feel like a travelling control room. I'm in touch via New Zealand with the family I'm hoping to fetch. They left Kyiv this morning, heading westwards, and they're now close to Lviv. People send me heart emojis, while Hungarian friends insist that the West has been goading Putin too much. For an English friend who works for an aid organization, I promise to find out what medicines are needed at the border.

I reach Záhony at 23.00 hours. It's deathly quiet. This is an old-fashioned Eastern Bloc border crossing: broad asphalt, bright lights, built to intimidate. The Ukrainian side of the border is hundreds of metres away, out of sight, on the other side of the River Tisza. On the Hungarian side, to the left of the road, their noses pointing westwards, a long line of cars is parked next to the crash barrier. They have Hungarian or Ukrainian license plates. Those sitting in the cars are mostly mothers. Children lie sleeping on the back seats. I ask what they're waiting for and they calmly wind down the windows, an undaunted look in their eyes. These people have something that many have lost: both feet planted in brutal reality. They are waiting for family members, parents, occasionally a husband or son. Ukrainian men aged between eighteen and sixty can leave the country only if they have 'special papers'. 'Even if they're the only breadwinner, or are travelling alone with children?' I ask. No, the women firmly shake their heads. 'In that case they have to hand their children over to a woman and drive back.'

A Hungarian police officer in a fur hat chases the parked cars away. I ask him about the length of the queue on the other side of the border. 'Four hours,' he tells me. A car crosses out of Ukraine once every fifteen minutes. I drive around for a while. It's almost midnight. From the Hungarian side three young men with small rucksacks walk towards the border, along the hard shoulder. As I approach they stick out their thumbs. I stop and ask what they want. They're just twenty years old, younger than my own sons. They tell me they want to go home to fight the Russians.

Kisvárda, Saturday morning, 26 February 2022, 09.00 hours

Sanctions

Like Napoleon Bonaparte, the management of the hotel in Kisvárda in eastern Hungary regards a substantial breakfast as vital for the troops. Only vegetarians get a raw deal here (thick slabs of bacon, black sausages, scrambled eggs, slices of ham, chunks of spicy *kolbász*, and as a finishing touch some red pepper and tomato). In the breakfast room I hear a mixture of Hungarian and Ukrainian. Practically all the cars in the hotel car park (lots of BMWs and Mercedes) have Ukrainian license plates. The drivers are as robust as their cars: people in practical clothing, tracksuits and trainers, unshaven men, lipsticked women. An older Ukrainian woman comes in, looking for her son. We're in the first decent-sized hotel this side of the border. The sweet, plump hotel owner speaks only Hungarian, the Ukrainian woman a little English. With my limited Hungarian I act as interpreter, and then the adult son walks in of his own accord. Mother in tears.

For the first time in my life I'm politically engaged. Zelensky's speeches touch me, and above all I admire the courage of ordinary Ukrainians: young mothers learning how to fire a Kalashnikov, saying they'll fight for their family, city and country; elderly women telling Russian soldiers the truth; and the heroic border guards on Snake Island who have made clear how they feel about the Russians in the plainest of terms.

While the Ukrainians fight to the death against a superior force, Western companies and politicians are unwilling to make sacrifices, especially the rich former Axis powers Italy, Germany and Austria. It's deeply shameful.

Putin must be dealt with, and the sanctions introduced must be more meaningful than exclusion from the Eurovision Song Contest. The time of half measures and 'avoiding provocation' is over. We need to supply arms, to impose sanctions that will bring the Russian population out onto the streets, to mount cyber attacks. We need serious reinforcements for the Baltic states and generous treatment for Ukrainian refugees. The old-fashioned logic behind what Putin is doing is alarming; he's trying to restore a cordon around Russia to make sure that the enemy is always so far from Moscow that if it attacks in the summer, it won't reach the Kremlin before winter and will therefore be defeated – that's what has always saved Russia, all the way from Napoleon to Hitler. One glance at the map of Europe and you know that after conquering Ukraine, Putin is sure to lay claim to the Baltic states, Moldova or Poland. The route from the Baltic countries to Moscow is, after all, a good deal shorter.

Meanwhile I've heard nothing from the Ukrainian family in flight. Contact is impossible. I don't know where they are and I still don't know which crossing they're making for. I'm sitting in a breakfast room with a gleaming granite floor, waiting while people fuel themselves up all around me.

Beregsurány, Saturday evening, 26 February 2022, 21.00 hours

Roma from Transcarpathia

It's all very moving. Women holding toddlers in their arms or by the hand are walking along the edge of the asphalt, out of the bright lights of the border post into the darkness of Hungary. Small children in thick Eskimo suits. They head off along the verge, away from the war. It's bitterly cold on the cheerless flatland where the Hungarian and Ukrainian border posts stand. There's a strange, semi-legal, smuggling atmosphere, with darkened cars parked on the dirt tracks and wasteland that surround the border, groups of people in the night, lots of blue flashing lights, police in high-viz jackets and an empty stretch of four-lane highway.

Five hundred metres from the border post, the Hungarian branch of the Red Cross has set up a tent. Behind it, on the edge of a stubble field, five blue plastic toilet cubicles stand at rakish angles. A young woman wraps a blanket around a child. She's a Ukrainian from Mukachevo (Munkács in Hungarian). It took her five hours to get across the border. Her husband and daughter are in Bratislava and they're coming to fetch her. A white canvas awning was put up by the local authorities yesterday and two men are tinkering with a couple of those mushroom-shaped gas heaters. Everything the village council in Beregsurány was holding in readiness for village fêtes has been brought out of storage. It's a refreshments stand of sorts, handing out coffee, bags of potato crisps, bottles of water and bars of chocolate. Some twenty men are crowded together. Food parcels

are being distributed. A thick-set Roma man with several women around him steps forward and points to the packages. A tall aid worker asks how many he needs. 'Forty,' the man says in Hungarian, and I realize at once how complicated it is to be an aid worker.

Refugee reception in Hungary is in the hands of the Maltese Red Cross. They're at the tent too, arranging for people with no one to fetch them to stay the night in Beregsurány, in the Kultúrház (Culture House) and the school. They're also organizing buses to Budapest, where further accommodation is being made ready. We've offered space at our house. A man from the Maltese aid organization tells me that the first stream of refugees were Roma from Transcarpathia (the western part of Ukraine, where a large Hungarian minority lives). Now the people crossing the border are mostly from Kyiv and the surrounding districts, to judge by the license plates on the cars that are making it through in dribs and drabs.

This war, just two days old, is making me a touch sentimental. I don't know why, but everything affects me, as if ancient fighting instincts have been released from under the numbing Teflon of decades of consumerism. The speeches by President Zelensky (at this point I don't know whether to spell his name with an i or a y. Names ending in y in Hungarian mostly belong to the old aristocracy. I've personally raised the president of Ukraine to the Hungarian nobility, whether he likes it or not), the women with infants who cross the border and speak in support of their husbands, the elderly ladies making Molotov cocktails who say this isn't the way they usually spend their Saturdays, the young men with puppy fat still in their cheeks who say passionately, and in beautiful English, that they want to take up arms and are willing to die. It's contagious.

 I've been to Ukraine a couple of times, but I've never looked at the country and asked myself where the best places would be from which to fight a guerrilla war. I mostly remember endless fields stretching from Kyiv to Odesa. Black earth. Vast plains with few places to hide. When I drove across it, the stubble was being burned off the fields: orange flames kilometres long in the pitch-dark night. The Russians will burn everything to the ground too, in their hunt for partisans.

Kisvárda, Sunday afternoon, 27 February 2022, 13.00 hours

Old treaties

Last night, in an act of nonsensical solidarity and magical thinking, I ordered *Kijevi jércemell*, chicken Kyiv. We all do what we can. The menu is in four languages: Hungarian, German, English and Ukrainian. It's lunchtime and I'm sitting in my hotel dining room, amid burly men who are putting away huge schnitzels and entire stews. I've just heard that the Ukrainian family on the other side of the border spent the night in the mountains and is on the road again. Shortly I'll try to work out which border crossing is quickest, but now I've got an hour or so in which to zoom out.

For the past two years I've been working on a non-fiction book about a ménage à trois involving three people and three empires, the story of a refugee from Estonia, a Dutch aristocrat with roots in Constantinople and a Hungarian peasant's son. All three lived at loggerheads for a time in a village in Hungary, close to my own village. The instigator of this impossible set-up was the Dutchwoman, a distant aunt of mine with an above average libido. All three were exponents of empires that had recently fallen: the Russia tsardom, the Ottoman sultanate and the Austro-Hungarian dual monarchy. Those three empires lived together for centuries, amid the inevitable domestic quarrels and conflicts.

Two of the empires imploded a century ago and new borders were drawn. After the First World War, the Ottoman Empire (in the Treaty of Sèvres, 1920) and Austria-Hungary (in the Treaty of Trianon, 1920) were stripped of much

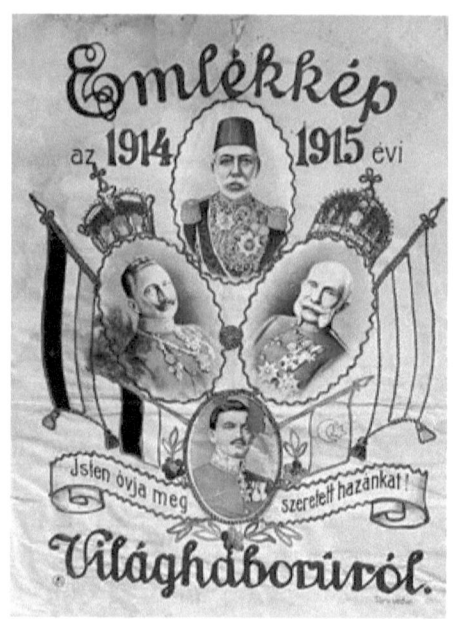

of their territory, while a handful of neighbouring countries were apportioned new lands. The Russian Empire experienced the humiliation of amputation seventy years later, following the collapse of the Soviet Union. For people in the peaceful, stable West it may be hard to imagine the extent to which revanchism can fester under the skin.

From working on my book – and occasionally following the news – I've gained the impression that those losses of a hundred years ago still shape this region. The former empires play their part to this day. Putin, Erdoğan and certain people in Budapest still dream of regaining lost lands, while Russia and Turkey are building strong armies to avenge old humiliations. I don't need to recall here all the aggressive acts carried out by Putin over the past twenty years. Erdoğan is still cautious in his territorial expansionism (in Syria). The large-scale disorder entailed by a war allows imperialist wet dreams to come within reach unexpectedly. So we can only hope this ends with the downfall of Putin.

After days of creeping along the border like a fox around a chicken coop, it seems I can finally make myself useful. The parents plus grandmother, a pet of an unspecified species and three young children were supposed to be coming my way this morning, before they decided on Moldova because the queues at the border were shorter. After that, Bulgaria was regarded as the best option and now, at the end of the day, they've settled on Hungary again. They're a five-hour drive away.

I immediately went to size up the situation at the Záhony border crossing. There were lines of buses with Hungarian license plates, freshly arrived from Ukraine full of people. The Hungarian government is intensifying its support. The wait on the other side of the border is now around twelve hours, Ludmila tells me. She's waiting on the hard shoulder on the Hungarian side, having flown in from London very early this morning to pick up her grandchildren. 'I'm Russian,' she tells me. 'But my two grandchildren are Ukrainian and I'm fetching them because they're being shot at by the Russians. My son and his wife are staying behind to defend the country.' She looks into my eyes and falls silent. She's wearing a fur hat with earflaps, like an old fur trapper, and trembling. Her nerves are in tatters. It's been a long day.

Kisvárda, Monday morning, 28 February 2022, 09.00 hours

Sergey Naryshkin

My phone has been pinging since yesterday evening, every thirty seconds, as if I've got a legion of mistresses. It's like this. Last night I made some Ukrainian friends in the hotel bar. In my own family I'm regarded as a Neanderthal when it comes to the digital world – and I consider it an honourable title. I can't stand those little phones, especially the tiny keyboards and the fact that everyone always knows where I am and what I'm saying. Three months back I started using a laptop and shortly before that, very occasionally, Google Maps, having relied entirely on paper maps until six months ago. Now, driving to the Ukrainian border, I've brought a stack of maps with me that cover Ukraine, Transylvania, Moldova and Russia, as well as a street atlas of Kyiv, although before leaving I had to make a solemn promise to my wife Ilona that I wouldn't go to Kyiv, or into Ukraine at all. I fear that if I hadn't made that promise, I'd have been unable to resist.

My new friends have helped me to get onto Telegram, and especially two messaging services used by Ukrainians. Now I'm receiving an unbroken stream of messages showing columns of burning Russian military vehicles and peptalks from a world champion boxer who says they'll fight to the death, followed by a torrent of hearts and biceps. It's great for morale. I think that if Putin wants to get his hands on Kyiv he'll have to use the same tactic as in Grozny (which is to say, bomb the city flat and have everyone he captures tortured, preferably by locally recruited accomplices). There's no way he'll manage that.

The great thing about the Ukrainians is that 'you can win a war with them'. They've shown us, with Zelensky out in front, that it's possible to be steadfast and principled, to stand for something.

The fact that we in Europe are now suddenly acting with such unanimity and resolve is down to the example that they're setting, that *he* is setting. I think it's also because we have so many women in charge in Europe these days (six female defence ministers), who hold the moral line more firmly than the haggling men. European countries are of one mind, showing their teeth for the first time, displaying fortitude. It's tremendous: send weapons; suspend Russia's Central Bank, so that all funds held abroad are unreachable; suspend the SWIFT payment system; and ban those two devious Russian propaganda broadcasters RT and Sputnik. When I heard that – I was in the car, on my way back from the border – I raised a clenched fist like a football supporter. Furthermore, oil giant BP is severing its links with Rosneft, under pressure from the British government. Couldn't the Dutch government, I wonder, lean a little harder on Shell and the Gasunie, so that they break all ties with Russia and Gazprom?

A week ago, on Monday, Putin held a consultation with his security council, a meeting in a huge room at the palace, its terrifyingly high ceiling covered in gold leaf, where his aides were ordered to come forward one by one to give their advice to the boss. They seemed to know instinctively what they were supposed to say. Only Sergey Naryshkin, director of the foreign intelligence service SVR, former KGB fellow student and friend of Putin, dared to speak honestly and propose something different, namely issuing a short-term ultimatum to the West. He was dismissed in front of the eyes of the world like a schoolboy: 'You can take your seat.'

That humiliation is mainly reported as a show of ultimate power by Putin, but I see it more positively. The KGB and its successor the FSB are held in high regard as organizations at the top of their field, with extremely intelligent staffs – superior to the CIA. Sergey Naryshkin is undoubtedly one of the best-informed people in the world when it comes to the benefits and drawbacks of a Russian invasion of Ukraine. He knows as well as anyone the possibilities and impossibilities, the low morale and drunkenness of the Russian troops, the strength of the Ukrainian armed forces, the availability or otherwise of fuel, the corruption among junior officers and so on. The fact that he of all people has doubts and proposed seeking a solution at the negotiating table is hugely encouraging. At the same time, it's disturbing, extremely disturbing, that Putin doesn't give a damn about advice from the head of his own foreign intelligence service.[*]

Now I need to get to the border. They're a hundred metres away!

---

[*] Later there were reports that Sergey Naryshkin may have received this reprimand because most of the weak points resulting in security leaks were suspected of being in his foreign intelligence service.

Lónya, Monday afternoon, 28 February 2022, 16.00 hours

Child's shoe

In the roadside verge at the border post in Lónya, a hundred metres from the final road barrier, a little girl's shoe lies in the grass, fluffy and small as a newborn rabbit. On the front of the shoe is a ribbon, with gleaming bits of plastic, like pearls, arranged in heart shapes on the loops of the bow. A little girl will have been inconsolable when she discovered, in a car or bus as it moved through the night, that the shoe was no longer on her foot. In the excitement and joy of having made it across the border, lugging all those belongings and children, no one will have noticed its loss. I've always been mystified by the large number of single shoes in motorway verges, but here, near the crossing at Lónya, I can see the accompanying drama before me.

I reached this small border post, in between the far bigger crossings at Beregsurány and Záhony, on a ferry,

which with a bit of inching and pinching had space for six cars. After crossing the border formed by the River Tisza, I found myself on bare land with a cold northerly wind from Siberia whipping across it. The road to the border is four metres wide, along a dyke set into the water meadows of the Tisza, a route for friendly smuggling. The narrow road winds through no man's land. In the distance to the east are the Carpathian Mountains, a handful of snow scattered over them like icing sugar.

On ordinary days this border post is no doubt used only by local farmers and the odd commuter or smuggler from Lónya. A few cars are waiting in the verge, along with a lot of police and soldiers from the special units. Broad-chested lads in camouflage radiate self-confidence and chat in the sun. A stressed-out police officer comes up to me shouting because I'm taking photos – as if we were in the old Soviet Union. She insists that I wipe them, which I do half-heartedly, deleting the blurred ones. I want to keep the picture of the child's shoe, naturally. She looks at the photos with me, which – as everyone knows these days – is not exactly polite. Fortunately, one of the men in camouflage comes over to us; from my height and my defective Hungarian he has recognized me as a Dutchman. He's often trained with them, with the Dutch special forces, and he speaks English. The fanatical police officer slinks away.

Between crush barriers, a queue of people about two hundred metres long stands in the biting cold. One window of the customs building is open, like the counter of a village railway station. I fear this overwhelming influx of refugees is being processed with customary rural efficiency and bureaucratic hair-splitting. I can't see the end of the line of cars on the narrow forest road between the Ukrainian and Hungarian border posts. The Ukrainian family is somewhere out there, between the two countries. Yesterday they sent the occasional pin to show their location. They're not far away, but there's no telling how long this might take. The soldier had no idea, a police officer I asked earlier didn't know either.

Past the child's shoe, which is still lying low, I walk back to my car. On the other side of the verge are two cars with a group of women and several children in them. I ask whether they speak English. They all do. They're waiting for a family member, and when she arrives they'll all head for Poland. Their husbands are in Ukraine doing what they have to do. I ask how long it took them to get from the Ukrainian border to this side of the Hungarian border post.

'Ten hours. We arrived at the border at six yesterday evening, got past the Ukrainian border post at one o'clock and came through into Hungary at eleven this morning.'

That tells me enough. I'll go back to the hotel for a bit. Waiting for ten hours in the north wind serves no useful purpose, it seems to me. I take the ferry back across the Tisza. The cable judders; the water flows grey-brown beneath us. We're halfway across when a group of Ukrainian children appears at the bend, above us on the dike. They whoop and scream with excitement at the sight of the ferry, a vessel attached to a steel cable, its deck lined with planks, that will take them to safety on the far bank.

Lónya, Monday evening, 28 February 2022, 23.00 hours

The border crossing at Lónya

It's only logical that every driver has to open up their car bonnet at Lónya and unload all the suitcases so they can be searched, now that Kalashnikovs are being handed out by the thousands in Ukraine with no questions asked. Hungary takes its task seriously; you don't want an unnecessary number of those easy weapons pouring into Belgium and Bosnia. Cars are coming through at a trickle. The women who get out from the driver's seat and struggle to fathom how to open the hood generally look extremely presentable. Wavy hair, light-coloured ski jackets, rather like the way my mother and her bridge friends looked twenty years ago. Middle-aged ladies with hair appointments at least once a month.

It's pitch dark when I arrive at the Lónya border post for the second time, not by boat but over a bridge and after an endless detour on narrow country roads, because the last ferry departs at five in the afternoon. Hurtling along winding lanes turns out to have been unnecessary, since the Ukrainian family I'm here to fetch is close by – a stone's throw away – but still far off in time. As before, a car is being let through about once every fifteen minutes. The flow of people on foot is comparatively brisk at this border crossing.

I spend hours at the tiny border post, and for the first time in my life I smell what a war is like and what fleeing for your life means. It grabs me by the throat. In the darkness I can't see the end of the trail of people crossing the border on foot. There are two queues, one for Ukrainians and citizens of the European Union and one for the rest of the world. The first queue takes an hour to pass the checks at both borders, the second four hours, which does make sense: Ukrainians and Europeans are allowed to enter the EU freely; everyone else has to apply for a visa.

The police and soldiers are so busy that I'm able to walk around undisturbed. The rest-of-the-world queue is behind crush barriers, like away-team fans at a football match, separated from followers of the home team who form a broad front beside them, neatly in line, in cheap thick coats, woolly hats pulled down over their ears, holding children's hands, their bags and suitcases on the asphalt. There are a lot of elderly people – it's often the grandparents who take children to safety. More and more buses are arriving, and they have barely enough space to turn round at the border post. There's something very ominous about commanding figures in uniform in a desolate landscape, loading huddles of exhausted civilians into stationary buses that leave their engines running.

I wander into the hall where visas are being issued. On the wall are colourful posters showing photos of the

birds of prey and reptiles that have their habitat in the surrounding area – a reminder of quieter times, when the only strangers who turned up here were ornithologists with binoculars hanging against their chests. Inside it's packed. Apart from two with a mild form of Down's syndrome and the police officers criss-crossing the space, I'm the only person with a Caucasian appearance. I ask a plump boy with glasses and an open face how long he's been waiting. 'Getting a visa will take another five minutes,' he says without looking at me. His eyes follow a police officer who is making his way through the crowd. The man goes into a room and comes out with a pile of passports so big he can barely clasp them all. He goes over to stand by the door, pulls himself up to his full height and shouts, 'Everybody for visum follow me!' Bags and suitcases are hastily repacked, and the space is sucked empty as if someone has thrown a thermobaric bomb. I'm left behind with the bird posters.

They all get into a bus. Outside there are shouts from between the crush barriers. Their turn has come. It's icy cold. The thermometer reads around freezing point, but the wind makes it feel more like minus ten. People are pressed together like animals. It's all extremely depressing. At first I couldn't understand why so many non-white people were pouring out of Ukraine, but in my hours at the border I learn a few things.

As soon as war broke out, the Poles shouted from the rooftops that everyone from Ukraine was welcome and they wouldn't be checking papers, they'd simply let everyone in. Poles and Ukrainians look a bit similar; they're closely related. That fact, along with the announcement, ensured that in the first few days practically all the refugees headed for the Polish border. The Poles had failed to realize that there were quite a few Indian, Nigerian and other African students in Ukraine, as well as a contingent of Afghan refugees who, because of vague promises by the Americans after the fall of Kabul, had been parked for the time being in amenable Ukraine. The Poles failed to come up with any constructive solution for this unexpected group of refugees, and at the border they kept sending them to the back of the queue as far as possible.

Hungary is checking papers as normal and then letting people through, so the students and others who have been waiting for days at the Polish border to no avail are now making for Hungary en masse. In the freezing cold, under a clear starry sky, in front of the visa office at the Lónya border crossing, it seems almost as if not Kyiv and Kharkiv but Lagos or Cairo is under bombardment.

Lónya, Monday evening, 28 February 2022, 00.00 hours

Gelsomina

The Ukrainian family I've come to fetch is approaching the gateway to Europe at a snail's pace. They've now been at this border crossing for fifteen hours. They're less than a hundred metres away from me, a young couple, three small children, a grandmother and a pet. I still don't know if it's a dog, a cat or a rattlesnake, only that they've got an animal with them. At the Záhony border crossing I asked a police officer whether pets were allowed over the border without a passport. He told me he thought it wouldn't be a problem in time of war.

The refugees stand outside the door of the customs building as if shipwrecked, blue with cold. I stand with those who've come to fetch someone. At the edge of our group a mother bends down over a child sitting on a suitcase. She wraps a blanket and then her own body around the child. She's wearing a thin coat. She shakes her shoulders back and forth against the cold and tries to keep the freezing child's spirits up. A sense of despondency strikes upwards from my icy feet.

Then a rusty white delivery van drives up and turns around next to the customs building, between us and the refugees. A young redheaded woman hops out and walks over to a police officer, talks to him for a while and then swings open the back doors of the van. She goes into a bus full of refugees with a cardboard box, walking along the aisle handing something to all the passengers. People look up happily. When she's finished she gets another box of surprises out of her van and does the rounds of the refugees, mostly African students, who are standing in the unbearable cold in front of the building.

She dances about like Gelsomina in Fellini's *La Strada*, with the same mournfully cheerful face. She pulls a stack of thermal blankets out of the van, goes over to the woman with the little boy, wraps a blanket around the child and another around the shoulders of the mother, who almost levitates with gratitude. Then she makes her way across to the Africans with her blankets. One of the police officers, who up to then has stood watching with his hands at his sides, decides to help her. Together they put blankets around the shoulders of the refugees who are standing shivering under the bright lights. At her urging another police officer decides out of the kindness of his heart to go with one of the people who have come to fetch someone and search the endless queue for three children. Roma from Transcarpathia stand in a circle, stamping rhythmically after having been given blankets by the woman. She strides about and speaks with refugees and the police.

From the collection of cardboard boxes in the van she manages to magic into existence all kinds of things that bring deliverance. The bitter cold and our powerlessness had almost defeated me and the washed-up souls on the Hungary-Ukraine border, but Gelsomina has saved us. She peddles charity from her dilapidated delivery van, transforming the mood just before midnight. It's as if a circus bringing comfort and joy has set up its tents at the Lónya border post with the ginger-haired girl in the role of ringmaster, magician and sawdust-sweeper.

Lónya, Tuesday morning, 1 March 2022, 00.30 hours

'When the bombs came, I took my babies'

For four days I've been staying in a kind of Little Ukraine – the hotel is packed with Ukrainians – in a room that's constantly heated to thirty degrees, with a knobless radiator, and with a voluptuous redhead above my bed courtesy of an amateur painter. Road maps in one hand, I've been roaming the border area like a novice smuggler, pointing to rivers and mountains, or asking dubious-looking Ukrainians in the breakfast room, local bar girls, aid agency staff and newly arrived refugees from Kyiv and Transcarpathia whether there's a spot where you can cross the border illegally. They all firmly shake their heads. I was once directed to a crossing where it was supposedly possible to bribe the customs staff, who would let you through for 300 euro per adult male. But my informant added, 'For all I know that whole team might have been replaced by now.'

Then, on Sunday morning, I heard that the rules in Ukraine are being relaxed. All healthy men aged between eighteen and sixty still have to stay in the country to fight, but there's an exemption for fathers with three or more young children (meaning those aged under eighteen). It's a number chosen for practical reasons; after all, a mother has only two hands. A mother duck can keep three young under control, while all the ducklings beyond that are eaten by pike or rats. A human mother can manage two.

This is good news. The father of the young family from Kyiv that I've come to fetch no longer has to do anything

complicated; they can all leave the country legally. They spent Saturday night with friends in a mountain hut in the Carpathians, trying to sleep for a few hours. Aside from that they've been living for days in their van.

At last Alexey and Yeva, her mother, their three sons Misha, Gosha and Jenya (aged eight, six and three) and cat Tasya cross the Hungarian border a little after midnight. It's Tuesday morning and it's taken them about twenty-four hours to get across. I stand thirty metres behind the border post, in the middle of a black asphalt square, in a knee-length, cobalt-blue woollen coat. The customs staff and refugees probably take me for an employee of the Swedish embassy or a pimp, since everyone around me is wearing tracksuits and trainers.

All this waiting, watching the endless stream of people at the border – the sunken eyes, the shoulders twisted by cold, the gaggles of people trudging through, at the end of their tether, with wheeled suitcases or sports bags, the martialling of chaos by police trying to appear self-confident, the impossibility of dealing with such an influx of people both humanely and according to the correct administrative procedures, before transporting them to a safe, warm place – it's heart-breaking. The Hungarian police, generally speaking, are strict but sympathetic.

With long strides and gestures, like a member of the ground staff at a primitive airport, I guide Alexey and Yeva's silver van to my parked car. The little boys are asleep in the back, folded over their grandmother, who smiles at me through the window, stiff with exhaustion. Only Alexey and Yeva can get out. Alexey slides from behind the wheel, which is horizontal, as in a big truck. He looks as if he's just spent a weekend on amphetamines: dilated pupils, unshaven and pale, wound tight by exhaustion. He starts rolling a cigarette and speaking in rapid Eastern European English. In telegram style he tells the story of their trip: 'When the bombs came, I took my babies and put them in the van.'

Yeva climbs out of the passenger seat, walks groggily to the roadside and comes to meet me at the back of my car. She has a beautiful oval face with a sensitive look. A broken person. They've made it. The whole family has got out of the country. She stands at the roadside, looks at me and whispers 'thank you' before bursting into tears.

Budapest, Tuesday morning, 1 March 2022, 03.30 hours

Bastard Putin

None of the family get into my car – the little boys are still lying asleep on the back seat. So I walk to the border post again. The girl with the delivery van, Gelsomina from *La Strada*, is still busily helping people. In Hungarian I tell her that I'm going to Budapest and have room for three people. Gelsomina immediately strides over to a big group of refugees standing in front of the customs building. It's a little after midnight. She pulls a couple of girls by the sleeve, mutters something in Ukrainian, turns round and asks whether I could take four.

'No problem,' I say. 'As long as they're not too fat.'

Four young women follow me to the car. I don't want to make the family wait too long, so I hurry to put the women's things in the back. Two have bulging suitcases with them, the other two only tiny rucksacks. While loading, standing next to the car in the pitch dark and not paying attention, I feel a passing car drive over the back of my foot.

Everybody who has stood waiting for eighteen hours at the border crossing is tired. The Ukrainian people are suffering enough already, so I don't want to draw attention to my heel. I can still walk. The women get in and we leave. The van carrying the Ukrainian family is already on its way to Budapest.

My foot is on fire. If I drag my left leg for the rest of my life, it'll be because of that bastard Putin. The advantage is that the searing pain keeps me awake until far beyond

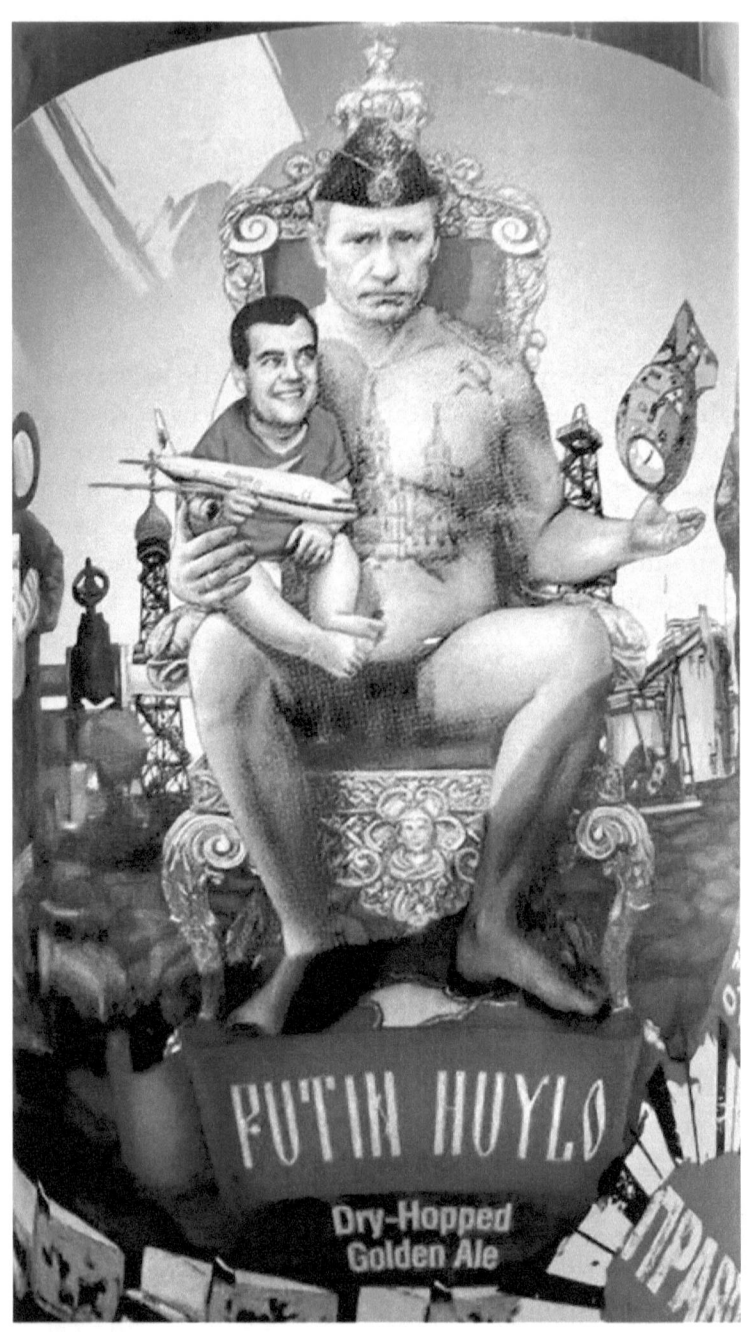

Debrecen. In the back, the girls chat urgently, in a whisper. It sounds like birdsong. They are shy. Only the one sitting directly behind me, out of my field of vision, lets out a cascade of words, in heavily accented English. The other three sit bolt upright, stiff with nerves, and say nothing. All four are from Kharkiv, the city in the east, close to the Russian border, that has been shelled and fought over for days. I ask what they do. The skinny girl in the middle says nothing at all for the whole journey. The one in the passenger seat – aquiline face, huge mass of hair – is studying computer science, the blonde woman behind her economics. A week ago they were attending lectures as normal. She's twenty years old and her answers are brief. It's not clear whether the women don't speak English, are introverts, are in shock, or prefer to leave the talking to the leader of the group.

'I'm Ella, I'm a barista,' says the redhead behind me. 'They're all students. I'm thirty-six and I only do simple work in a coffee bar. These villages are just like my village in eastern Ukraine, only the roads are much, much better.'

Ella knows the students as customers at the coffee bar where she works. They all fled Kharkiv separately after the first rockets hit. The four of them happened upon each other along the way, somewhere on the 1,400-kilometre journey from Kharkhiv to the border, in trains, on station platforms and in air-raid shelters, so they formed a group with the barista as leader.

The girl at the back right, the blonde one, the prettiest of the four, turns on the torch on her phone. She concentratedly applies lipstick, then rouge and mascara. A tiny rucksack is all she has with her, but there's no lack of make up.

'Her boyfriend lives in Budapest; he'll be there to fetch her,' Ella says, laughing loudly. 'She has to get ready.'

As for Ella, she wants to travel on from Budapest to Turkey, where she worked in the tourist industry before Covid forced her to return to her native city two years ago.

There she learned to make good coffee. One of the students wants to go to Vilnius in Lithuania, where her sister lives. What the silent one wants I've no idea. The women must be done in after days of travelling, waiting, sheltering, uncertainty. We drive along winding B-roads. I've switched the headlights to full beam.

Three deer step onto the tarmac, so I brake. In the light of the headlights they stride jauntily along the road, surreal, directly ahead of us. Their black noses tremble. We stare in silence for several moments at the deer. It's like a dream. They're very light in colour. The one at the front, the leading female, stands on the white line, one front foot in the air. She hesitates, then turns round. She doesn't dare cross over and at a trot all three go back to the side they came from.

A little later, at Nyíregyháza, we turn onto the M3 and I get up to full speed. I want to reach Budapest as quickly as possible. I'm worn out. Ella is cheerful again about the quality of the road – the black tarmac of the M3 lies broad and empty as an airport runway before us – and remarks that it's just as if we're in a plane and about to take off.

Apart from the girl with the Ukrainian boyfriend in Budapest, they have nowhere to sleep. I ring Will Clothier,

owner of the Brody Hotel and Brody House. If they haven't rented out all their rooms to an American film producer, they might have places for the night. A little later it turns out he has three beds. We can come. Shortly after that all four are lying asleep in the car like broken buttercups.

Aside from a few vans and some cars with Ukrainian license plates, the motorway is deserted. It's unreal, whizzing through the night with four sleeping girls. What kind of journey has it been for them? What have they left behind? From time to time the light on one of their phones flickers on and they take a look, before going back to sleep. They grasp their phones in both hands – it's all they have to hold onto in a time like this. While the others sleep, Ella tells me she has a son. She left him behind with her ex-husband. It was better that way. She repeats it several times. I hold a bit of chocolate over my shoulder, but she doesn't want it. She's got a knot in her stomach. She hasn't had a bite to eat since Thursday, since the start of the war, six days ago; she finds it an effort even to drink.

At about three I drop them at Brody House, where Will is waiting. You can win a war with a man like that. I'm embraced as I say goodbye to the women. They go up a beautiful wide, winding, dual-monarchy-era staircase, carrying their luggage. I hear Ella call out cheerfully from high in the stairwell, 'We're sleeping in a castle!'

Brody House, the boutique hotel, is on Bródy Sándor utca, diagonally opposite the radio station where on 23 October 1956 the shooting began in the Hungarian Revolution, when the Hungarian secret police, trained by the Soviets, opened fire from high in the building on a crowd of peaceful students down in the street who were demanding access so they could broadcast their key policy demands. Hungarian soldiers on duty nearby tore the red Soviet stars off their uniforms and sided with the students. The way the radio building looked after the fighting, shot to pieces, is the way all of Kharkiv looks now.

Somogy, Saturday 5 March 2022

Adopt a roadblock

The only friend of mine who was still in Kyiv, Dirck, not a Ukrainian but a Dutch citizen and therefore not duty-bound to defend the country, left three days ago, heading west. Dirck runs a big arable farm to the west of Kyiv. He's now at the border with Romania. I've heard from various people that getting across the Hungarian border is a much quicker process now than in the first few days of the war, when I was there. 'We haven't had any serious problems along the way,' Dirck writes. 'But we passed innumerable heavily guarded roadblocks, every one of them manned by dozens of armed volunteers. I don't see how Putin can ever conquer this country.'

I ask whether he's brought his hunting rifles and ammunition from Kyiv. He tells me he left them at home, but that in the countryside he passed roadblocks set up by local huntsmen, armed with double-barrelled shotguns and hunting rifles. I suggested that with the hunting fraternity in our Hungarian village we could take responsibility for a checkpoint. He apped back enthusiastically that we should start an international initiative: 'Adopt a roadblock.'

Ever since Zelensky announced that foreign fighters are welcome in Ukraine, I've had to stop several of my friends in Budapest from answering the call. It's hard to resist; the courage of the Ukrainians is contagious. No doubt a good many unmarried soldiers of fortune – Americans, South Africans, Brits and members of the French Foreign

Legion, who because of Covid were at home twiddling their thumbs – will have travelled to Kyiv and environs by now. Some 38,000 women are in the Ukrainian army, including a former Miss Ukraine.

Dirck apps from the car as he drives through western Ukraine: 'It snowed last night. As a hunter I know that your targets are more visible then. I think the snow will help the Ukrainian troops. The country is deeply moved by the support it's getting from all over the world. It's clear that no matter how this ends, our world will never be the same again, unless someone very quickly puts a bullet through the head of that criminal in the Kremlin.'

Somogy, Sunday 6 March 2022

'Russkiy voenny korabl idi na huy!'

The sixty-kilometre Russian military column from the north, coming out of Belarus, is not getting any closer to Kyiv, simply because the troops are refusing to go any further. The officers roar and rage, but the ordinary soldiers who've come from Belarus say, 'There's no way we'll fight our Ukrainian brothers.' This is part of a widespread problem. Putin and his generals are now hard at work bringing the troops together and even, in a clear sign of desperation, mobilizing riot police from Siberia, Kirgizstan or Uzbekistan.

That too is reminiscent of moves made during the Hungarian Uprising of 1956. When the revolt broke out, Soviet Soldiers refused, or at least hesitated, to open fire on the great mass of Hungarians, most of whom were unarmed, or armed only with Molotov cocktails and hunting rifles. After the members of the Hungarian delegation that went to negotiate with the Russians were all arrested (and later hanged – lesson: never send your top people to negotiate with the Russians), tanks were deployed in overwhelming numbers to take Budapest, manned by troops from far-flung parts of the Russian Empire, soldiers with an Asiatic background, to reduce the chances of them showing any mercy. In Ukraine, Putin cannot make use of merciless tank crews, because the tank columns have been fired upon from left and right for a week. He's resorting to increasingly atrocious ways of making war and will, I'm afraid, continue to do so; he gives the impression of being a bad loser. Yesterday I tried to buy iodine pills in Budapest. The pharmacist told me they'd sold out all over Hungary.

The Russian troops in Ukraine are disorientated. The road signs around Kyiv have been removed or changed. In some places the firm in charge of Ukraine's signage has switched place names for words that will go down in history, the answer given by the thirteen coastguards on Snake Island in the Black Sea when a Russian warship ordered them to surrender: '*Russkiy voenny korabl idi na huy!*' 'Russian warship go fuck yourself' (My Ukrainian friends tell me it actually means something even more spicey.) That heroic text has now appeared in Russian on traffic signs in Ukraine. Where once several place names were indicated, you now read 'Go fuck yourself' and underneath, 'Go fuck yourself back into Russia.'

The demoralization of the Russian troops is taking effect, as demonstrated by the many video clips of fleeing and captured Russian soldiers that I see in the Ukrainian chat groups. Dirck has sent me photos of billboards saying something like 'Russian soldier: better a court martial than going home in a body bag.' There are also a lot of billboards that read 'Putin is a prick', which in Ukrainian is more poetic and less banal. I keep hearing Ukrainian around me (we have two families in our house in the hills of Somogy, with a total of five children) and the stress on the vowels makes it sparkle. I ask Alexey to teach me the now legendary sentence *'Russkiy voenny korabl idi na huy!'*

The road signs need to be replaced all over Ukraine, Alexey says, so that the Russians won't know where they are. Weeping with laughter, I sit by the open hearth with Alexey, Yeva and Ilona. Alexey has just shown us a film of a villager on Telegram that makes clear how the Russians are being defeated by peasant cunning and ingenuity. An old man at a crossroads can't think how to remove the place name painted onto the asphalt and decides to crouch over it. He pulls down his trousers, takes a shit, and smears out the crap with his hand until the writing becomes illegible.

Somogy, Wednesday morning, 9 March 2022

Mamas in Europe

'Who knows the best divorce lawyer in Odesa?' was one of the more dramatic questions on Mamas before 24 February 2022. Mamas is a Ukrainian chat group where mothers, mostly young mothers, give each other help and advice. It has almost 25,000 members, and in the war it's turned out to be the quickest and best information channel for finding out where rockets are hitting, where you need to dash to the bomb shelter, or where food and medicines can be found. Mamas communicates far more quickly than the ministry of defence or other Ukrainian government institutions. It's about five hours ahead of the official news.

When Covid-19 overran the world, a separate Mamas group on the subject was started. Now that Putin is chasing Ukrainian mothers and children out of their country, a dedicated 'Mamas in Europe' chat group is up and running. It has 1,500 members and is growing rapidly. 'Mamas in Europe' features questions about documents, and information about where it's still possible to buy fuel for your car, which border posts are quickest, or where you can find a paediatrician in the neighbourhood of Rheims.

The smartphone has become an important player in this war. The Russian soldiers who were sent to Ukraine had to hand in their phones, probably to avoid the home front finding out what was going on. These were young people who were at a loss without their smartphones. The Ukrainian soldiers were simply allowed to keep theirs, and they filmed themselves blowing up Russian tanks, leading lines of Russian prisoners of war dressed in rags, and engaging in other morale-boosting activities. The Ukrainian population is filming enemy troop movements on a vast scale from flats and farms, and using pins to indicate the position of Russian columns so that a Bayraktar (the Turkish drone) can be sent to destroy them.

Украинских матерей лучше не злить

Everybody, to an equal degree, whether a civilian, soldier or refugee, clutches a phone to their bosom. It's the key to knowing what's happening, where everyone is and how your loved ones are doing. Yeva tells me it's the first thing she does every morning: look to see whether her brother, father and other people she loves are still alive. The journey from Kyiv to Lónya took four nerve-wracking days, which Yeva spent staring wide-eyed at her phone while Alexey drove, her mother kept the three children quiet and the boys calmed the cat, and vice versa. Yeva and Alexey's family was able to weave its way between the falling bombs and collapsed bridges with the help of the Mamas chat group. They drove from one point to the next, zigzagging through the provinces of Zhytomyr, Khmelnytskyi, Ternopil, Ivano-Frankivsk and Zakarpattia. They headed for Moldova for a while, because that was the best place to smuggle men out of the country, across the river, for 1,000 euro a head. A column of Porsches, Maseratis and Range Rovers dropped men off there. The women and children crossed the border legally, the men at night, in a small boat on the black water. Because of the new regulations, which mean that fathers of three young children are officially allowed to leave, Alexey didn't have to get involved with that dubious business.

'I'd used Mamas before, when we were looking for an apartment in Kyiv,' Yeva tells me. 'But now I was using it to find out where rockets were hitting, and which towns and villages we had to avoid. I've been on it for six years, since Gosha was born. The first night of the war, the night of 24–25 February, I couldn't sleep and I was on Mamas nonstop. We were staying with Alexey's parents in a suburb of Kyiv.

I got updates every few seconds. The mothers of Ukraine were awake all night like me, and they wrote where the explosions were and how many there had been – hours before it was on the news. It's because of Mamas that I understood we had to leave. We got up at six, packed and left.'

Their three little wild boys – Misha, Gosha and Jenya – romp around with our dogs, play disorganized floor hockey and regularly run inside, cheeks red with excitement and covered in mud, to nab a biscuit or a slice of cake. Partly thanks to Mamas, the future forward line of the Ukrainian ice hockey team has ended up in Hungary.

Somogy, Thursday afternoon, 10 March 2022, 16.00 hours

A cat called Tasya

On 24 February, in the space of twenty minutes, Yeva and Alexey grabbed from the cupboards in their apartment in the Karavaievi Dachi district of Kyiv everything they thought they would need for their escape. Yeva stood in front of her three sons' wardrobe ramming clothes into bags. The air-raid siren sounded endlessly, which didn't help her to concentrate on packing for a journey of unknown duration. When she opened the bags in Hungary, she discovered they'd stuffed mainly T-shirts into them, no trousers or jumpers – 'What was I thinking? It's winter!' – and only underpants belonging to Misha, the oldest of the three.

A debate arose as to whether to take the cat Tasya with them. They could put out food for three days and change the litter tray, but it occurred to them they couldn't be sure they'd be back within that time. So Tasya, in a grey plastic cat basket with a grid at the front that enabled her to see something of the world, was lugged down twenty-one flights of stairs through the echoing stairwell. In the silver van, Tasya rode to Hungary in the close company of the three sons and the grandmother.

Yeva tells me that a year and a half ago, Misha started insisting they get a dog. On the way to school she showed him that a dog needs to be taken for a walk in the park every morning, early, before lessons, then again when you got out of school and again later in the evening, even in winter. At that Misha observed, 'A cat is nice too.' And so, a year ago, Tasya joined the family, a marmalade cat with a white bib and white paws.

In the no man's land between the Ukrainian and Hungarian borders, Tasya escaped from the silver VW van and disappeared into inhospitable thorny shrubbery. Yeva crawled after it as far as she could go. On hands and knees under the thorn bushes, she begged Tasya to come back into the arms of the family, which after an hour and a half the headstrong creature decided to do.

Even now that they've arrived at our peaceful house in the remote village in Somogy, big-city cat Tasya needs to keep a low profile and stay indoors, since our dogs get endless fun out of chasing cats. At the stables and around the shed that stores food for the carp ponds, we have a swarm of barnyard cats to chase away mice and rats. Our dogs can't get hold of those cats, they're wild and savage, not the least bit impressed by stone martens, foxes or jackals, let alone a bunch of spoilt dogs who sleep every night in the warmth of the house.

Yesterday Yeva and Alexey took Tasya to the vet in Kaposvár to get her injections. I rang Dr Gippert's veterinary practice, where we always take our dogs. They could turn up without an appointment. When they tried to explain to the receptionist, in English, why they'd come, one of the vets walking past recognized the Ukrainian accent. The vet comes from Transcarpathia originally and he treated Tasya, giving her inoculations and a chip, free of charge. Tasya became the first of the family to be issued with an EU passport: ultramarine with a circle of gold stars, like a floating halo.

Somogy, Friday evening, 11 March, 22.00 hours

Where to find a father like that

'There's fighting now in the town where my father lives. More and more Russian tanks are arriving all the time. They're coming from the north and east,' says Yeva. 'The closest city, Chernihiv, has been completely shot to pieces. All the beautiful towns around Kyiv are being destroyed, razed to the ground: Brovary, Hostomel, Bucha, Irpin. My father can't leave, he doesn't want to leave, he's not going to leave. There's nothing further to be done for his wife, no more chemotherapy, she can't go to the hospital anymore. She barely gets out of bed these days. He's looking after her, and he'll stay with her however fierce the fighting gets. She got cancer in Chernobyl.'

The tension is written all over Yeva's face. She rings her father several times a day. Four years ago he moved to a little town north-east of Kyiv with his wife, for her sake. It was great to leave Kyiv; the woods and the outdoor air would do her good. He's sixty-one, owns a house with a piece of land, keeps chickens and goats and is consistently positive on the phone, Yeva tells me. He always makes jokes.

It's Yeva's father's second marriage. Yeva's mother is with a man who is caring for his granddaughters, because his son is a jerk who refuses to take responsibility for his own offspring. I recognize from Hungary the close intergenerational relationships and the active role of grandparents in the lives of their grandchildren. People often live in three-generation households here. The grandparents look after the children while the parents are at work. Yeva's mother's second husband lives in the south-east of the country, near the nuclear power plant at Zaporizhzhia, which is being shelled by the Russians. He wants to leave; he's got a car, but he can no longer drive. He's looking for a driver who can get him and his granddaughters out of that hell.

'Completely organic,' Yeva says of her father. 'He makes his own goat's cheese. And grows a lot of cucumbers. The boys love going to stay there. They always search for mushrooms with their grandfather. Every year, in September, as soon as it's rained, we hurry off to see him. They go into the woods with him and come back with big white mushrooms.'

Yeva shakes her head affectionately, her eyes shining.

'What's the village called?' I ask.

'Kozelets,' Yeva tells me. 'It's halfway between Chernihiv and Kyiv, seventy kilometres from Kyiv, close to the highway, the E95. The road the Russians advanced along. There's fighting. There are tanks a hundred metres from my father's house.'

'Fierce fighting is going on around Chernihiv,' Alexey adds. He's the spitting image of soccer star Zlatan Ibrahimović, both in face and body – big, muscular, fit – and temperamentally: full of fire, hot-headed, stressed out by worry, rather dominant, but also warm, open-hearted and caring.

'Our fighters destroyed a long column of Russian tanks there and shot down a Russian jet. The Russians couldn't take the city; it's a fortress. They're now going around it. The only thing they can do is bomb it and shell it from a distance. They're surrounding the place.'

From inside comes the sound of children crying. Yeva stands up and goes into the guesthouse. Alexey pokes at the fire and throws more wood on it. I look up Kozelets on my phone. It's an hour and a half's drive from Kyiv. The town's Wikipedia page shows a white church that has towers with turquoise roofs. It looks familiar. I start reading about the little town and realize that I've been there, long ago. It's one of the few Ukrainian places I've visited. The cathedral was dilapidated then, yellowish in colour, with flaking stucco. It was for the cathedral that I'd come.

A friend, Tatjana Razumovsky, had told me about Kozelets. Her ancestors financed the building of the cathedral. Tsarina Elisabeth, on a pilgrimage to Kyiv, was once stranded there because the coach carrying her needed repairs. It took hours, or days. A tsarina can't stay in a village inn, so she spent the time in the town's cathedral. The local boys' choir was drummed up. One of Tatjana's ancestors sang in it. He was an attractive boy with a velvet voice, which didn't pass without notice.

'We haven't got victory in a battle or a war to thank for our title, like most families,' said Tatjana, smiling broadly. 'But instead services between the sheets rendered by one of my forefathers.'

Tatjana was being extremely modest about her family's importance. Another ancestor was among those who, along with Catharine the Great, mounted the coup that deposed the disastrous Tsar Peter III in June 1762 and had him strangled by an army officer a week later, which put Catharine on the throne. For god's sake let that coup, in which an idiot was deposed 260 years ago and a woman who read Cicero, Tacitus and Voltaire took his place, be inspiration for Spetsnaz teams, FSB officers or, even better, a mistress of a tormented Putin. A snippet of polonium in his glass of champagne or in the condom that will inevitably be provided: if someone is thus inspired, I'll immediately travel to Kozelets and, in a cathedral financed by Tatjana's family and dedicated to the Holy Virgin, I'll light a candle for the hero or heroine.

Yeva comes back from the guesthouse and sits down by the fire again. 'My father is dedicating himself to Polish-Ukrainian friendship and collaboration.'

I hear her begging, praying that he leaves the city in time. His intensive contact with Poles will not be taken lightly by the Russians and they'll discover it as soon as they check his phone or computer.

'You know, the boys are extremely fond of my father; he looks like Father Christmas, with a big white beard. He goes out to pick berries with them, he knows a lot about nature, he used to be a hunter. He always took me with him, too. I've got photos of myself with my father, as a toddler with a gun, bandoleers of live ammunition over my shoulder and a dead rabbit in my hands. I'm the oldest. He'd have liked to have a son as his firstborn, so I always got to go out hunting with him.'

She smiles and shakes her head again, then stares up into the starry night sky above Somogy. It reminds me of the beautiful sentence that ends the story 'Crossing the River Zbrucz' in *Red Cavalry* by Isaac Babel: "'And now I want you to tell me,' the woman suddenly said with terrible force, 'I want you to tell me where one could find another father like my father in all the world!'"

Budapest, Saturday afternoon, 12 March 2022, 18.00 hours

Killing of the dream

The Ukrainian people are proud of the biggest aircraft in the world, a transport plane that resembles the Spruce Goose built by Howard Hughes. The Antonov An-225 was christened Mriia by the Ukrainians, meaning 'the Dream'. It was built in the 1980s, with a wingspan of 190 metres, and unlike the Spruce Goose the thing could actually fly. One use of the Dream was to transport medical supplies. It was at Hostomel airfield on 24 February 2022 for maintenance work on one of its six engines.

On the first day of the war, three weeks ago, Russia started to bomb and clip the wings of military installations – radar, airfields, communications hubs, ammunition depots – including Hostomel airport. It was chosen as a bridgehead and occupied by elite Russian troops, flown in by helicopter. The An-225, Mriia, was destroyed by the Russians as a symbolic act. Ukrainian special units soon managed to retake the airport. President Zelensky reacted in a broadcast message by saying, 'The Russians can destroy our airfields and aircraft, but they can never destroy our dream.'

Some people are convinced that Putin went mad after two years in Covid isolation, which he spent studying eighteenth-century maps of tsarist Russia in the Kremlin library. In the early days of the invasion, the BBC called it Putin's 'Covid project', comparing it to the way others spent their time confined at home during the pandemic tidying the attic or building a kitchen extension. Which doesn't detract from the fact that I think Putin is still his old calculating self, a streetfighter with a knife in his pocket.

From messages from people who have studied the matter, I understand that the Russian army is ill prepared, that its logistics are abominable, that at the start of the war, when the ground was still frozen, it didn't push on far enough and the vast expanse of fertile black soil, the chernozem, is now thawing and turning into endless lakes of mud, so that military columns and tanks are condemned to take the asphalt roads. And that the equipment and food for the vast majority of Russian troops are wretched, that the ordinary soldiers barely know where they are, that most of the pilots have no more than 100 flying hours (that's nothing; my wife Ilona, a hobby pilot not called upon to engage in aerial combat, has 300; amateur pilots in Western Europe can barely get insurance with less than 150 flying hours), and that dress uniforms were piled up in the backs of trucks and armoured personnel carriers, because it was thought that after two days there would be victory parades in polished boots on Taras Shevchenko Boulevard and along the Khreshchatyk, the main street through Kyiv.

The obvious explanation for this optimism and logistical idiocy is that Vladimir Putin thought up and prepared for this war in the cellars of the Kremlin with a small group of insiders. The ongoing conflict is now costing him hundreds of millions of euro a day. He can't keep that up forever. Pictures of hungry Russian soldiers plundering village shops and chasing after chickens in Ukrainian chicken coops with sticks are being shared in large quantities on Telegram. The fact that Putin is now recruiting soldiers in Syria is a sign that he's growing desperate. What does he expect from them, for god's sake? They'll shiver with cold, be discriminated against by Russian soldiers and run off at the sight of their first Ukrainian. Yeva tells me that her former neighbour is lying low in Donetsk – hidden by his wife

– because in separatist-ruled areas all the men are being loaded into buses to go and fight for the Russians in Kharkiv. They're taken out of factories or picked up on the streets, given shabby uniforms and sent off to serve as cannon fodder. Putin will have to hurry up and drive Zelensky, the Ukrainians or the West to the point where they throw in the towel. What has he got up his sleeve for achieving that?

In Mariupol people are now drinking water out of radiators and have been instructed by the municipal council to tie the arms and legs of the dead together and lay them outside. Where do we draw the line? When will we decide that we don't want to help to finance Putin's war any longer in any way? When will we be prepared to stop using Russian gas? After an attack on twenty hospitals? After hundreds of thousands of people in besieged cities have starved to death? After chemical weapons are used in the middle of a city? Or only after a nuclear attack? How many more dreams need to be destroyed before we're prepared to suffer from cold or reopen the Groningen gas fields and generously compensate anyone who is affected by the resulting earth tremors?

Somogy, Sunday evening, 13 March 2022, 19.00 hours

The story of a chicken

At the point when the war began, Alexey and Yeva were as far apart as it's possible to be in Ukraine. He admits that he found the prospect of all out war absurd, unthinkable. Yeva was lying in a hotel bed in Kharkiv and Alexey was in Odesa. Yeva's mother was looking after their sons in their flat in Kyiv, on the northern edge of the city. At four in the morning the bombing of military targets in and around all three cities began.

Yeva is a former photo model. Thirty-eight years old, she has a sensitive face, and as you'll know by now she's the mother of sons aged eight, six and three. She spent ten years travelling the world for photo shoots and fashion shows, then trained in photography in Milan and worked as a professional photographer (mainly fashion and portraits). Since becoming a mother, she has coached young photographers and given online and offline masterclasses. On the morning of the twenty-fourth she leapt out of her hotel bed, threw her things into her suitcase and rushed to get on a train. 'We were fortunate that the hotel was close to the railway station. It was impossible to get a taxi and no buses or trams were running.'

She ran through the city with her suitcase and just managed to catch the packed seven o'clock train to Kyiv. Whenever she could – for whole chunks of the five-hour train journey from Kharkiv to Kyiv there was no signal – she rang her mother and asked how the boys were doing, whether they were still alive. Her mother was perfectly calm: 'Everything's fine. We're just having breakfast.'

'I kept being unable to contact her and after twenty minutes I'd go crazy,' Yeva tells me. 'I knew that Kyiv was being bombed. I tried to meditate in the train to calm myself down, but I couldn't.'

Alexey was woken by explosions too. He'd been in Odesa for weeks, installing an air traffic control system at the airport. It was almost finished, with just a day's work to go. Alexey is a network engineer, a graduate of the Technical University in Kyiv. He took the company's VW transporter van and set off in the early morning to drive like the wind to Kyiv, five hundred kilometres northwards through a country under fire. In Yeva's case the maternal instinct was dominant; she didn't take any photos of the crowded train or the scenes around her. Alexey, with the analytical brain of an engineer, took one-handed photos with his phone as he drove, of the plumes of smoke he saw to the left and right and of the endless lines of cars leaving Kyiv. He was one of the few lunatics trying to get into the city.

He arrived in Kyiv just in time to collect Yeva from the station, and a little later they ran together up the endless stairs to their flat on the twenty-first floor in the Karavaieva Dachi district to find their sons in the care of babushka, Yeva's mother. They hugged the children tight and packed their things.

In the midst of the blind panic that takes hold of you when your city is being bombed and you need to get your three small children to a place of safety, babushka was calmness personified. As indeed she still was in Budapest, where the whole family stayed for several days, and is now in Somogy, in our guesthouse. In the background the mother and grandmother is always there, quiet and with a loving smile, cleaning, cooking, folding clothes, taking her grandsons onto her lap and stroking their hair, like the sun around which the other planets turn in their own chaotic orbits.

Alexey and Yeva threaded their way through the shattered city in the silver VW van. On that first day of the war, Kyiv was like an antheap that somebody had poked with a stick. They were on their way to Alexey's parents, who live in a suburb to the south of Kyiv. They would sleep there that night and the next day head for Lviv at last. They were already quite a long way south, on Chervonozoryany Avenue, when Yeva exclaimed, 'The chicken!'

It's three weeks later. We're sitting at the open fire by the guesthouse in the countryside of Somogy where Alexey, Yeva and family are able to catch their breath for a while, when they tell me the story. There is loud laughter. That first day of the war they turned round and went back to throw away the chicken they'd left in the kitchen. The city was so congested that the whole exercise, driving there and back, running up the stairs and down again, took them an hour and a half.

'When a missile might hit at any moment, when you're sitting in a van with your whole family, trying to get out of a city that's under bombardment, you go back to throw away a dead chicken?!' I ask.

'I was worried the flat would stink when we got back,' says Alexey.

'Was it a whole chicken or chicken breasts?'

'A whole chicken.'

'Plucked?'

'Yes, ready for cooking. I've got one of those little barbecues, egg-shaped, a Kamado grill, on the balcony. I'd asked babushka to take the chicken out of the freezer. I was hoping to get home that Friday evening after weeks away and I wanted to celebrate with barbecued chicken.'

'The chicken was on the table?'

'On the countertop.'

'And you threw it into the bin indoors or outdoors?'

'Outside, of course.'

The story of the chicken illustrates the idiotic reality of war, and indeed the power of denial and above all the bound-

less optimism of the human species. Alexey thought he'd be back in his flat within days, or a few weeks at the most.

Diagonally opposite the guesthouse where Alexey and Yeva are staying is the rural retreat of my brother-in-law and his wife. Her family fled Russia via the Crimea, ending up in Paris, and thought they'd be able to return to Moscow within a few weeks. That was in October 1917. They've never been back.

A little further away, near the church, is my in-laws' holiday home. My father-in-law fought the Russians in the Hungarian Uprising and was forced to flee the country when Russian tanks took the city by brute force. That was in November 1956. Ilona and I were the first of the family to return and settle permanently in Hungary, in 2003, almost half a century later. I fear that Alexey, Yeva and their three sons Misha, Gosha and Jenya won't be able to go home until long after the maggots have finished their work.

Somogy, Monday afternoon, 14 March 2022, 16.00 hours

The dramaturgy of a mass murderer

Vladimir Putin, as we all know, is a product of the Soviet Union, having grown up in a damaged family in a damaged city in a damaged country. He was trained by the KGB, and from the start of his career he surrounded himself with KGB friends – he *is* the KGB. In some people this may evoke romantic associations with leggy women in mink coats, Koshi Waza and James Bond, but the reality is obscene. The FSB is all about psychological warfare, intimidation, blackmail, murder, torture and terror.

To get an idea of what awaits us, and above all the Ukrainians, you need to look at Chechnya and Syria – and at Dzerzhinsky. What happens in Russia and is planned for Ukraine can be traced back to Felix Dzerzhinsky, who developed the methodology of terror for Lenin. Dzerzhinsky was a Pole born in nearby Minsk, capital of present-day Belarus. He was responsible for the Red Terror, the arrest and execution, within a day, of thousands of class enemies in the cellars of Moscow's notorious prison the Lubyanka and elsewhere. He set up the secret service, the Cheka, which has since been transformed via the NKVD and KGB into today's FSB. Felix Dzerzhinsky is therefore the father of the KGB mentality: 'We represent in ourselves organized terror – this must be said very clearly.' Dzerzhinsky is the seed, Putin the harvest.

After its occupation of the Eastern Bloc countries, the Soviet Union had roughly the same problem as the Russians will have in Ukraine, if Putin manages to occupy all or part of the country: how to subdue the population. He'll be forced – and I'll stick my hand in the fire if I'm wrong – to deal with it using the methods devised by Dzerzhinsky. The Netherlands, and even more so the criminal Germany, will have to shut off Russian gas, if only because of the unfathomable repression that will follow a Russian victory. What do they think is happening at this very moment to the abducted Ukrainian mayors and other resistance leaders? If the interrogators are content with stubbing out cigarettes on their lips or giving electric shocks to their balls we should be relieved.

In 1950 the Russians set up the Dzerzhinsky Academy to train aspiring members of the AVO (the Hungarian secret police under communism) in the methods of terror. Some time ago I spoke to a Hungarian historian, Zsófia, who specializes in the subject.

'At that academy people were taught how to build a file, how to prepare a court case, how to tackle the dramaturgy of a trial,' she told me.

Dramaturgy – it's not a word you immediately associate with the Russian president, but he's had lessons in it, and suddenly you recognize the fact. Putin may not be much good at logistics, but he's no slouch at dramaturgy, or at set dressing. Everything we get to see has been thought through, so that we see it the way we're intended to. Think of that uneasy meeting with his security advisers on Monday 21 February 2022, in that tall room with its colonnade of pillars. Or the reception given to President Macron at an immensely long white oval table, or the consultation with his small group of close aides, at a far longer table still. The decor and the choreography are brilliant every time: Putin as the ultimate villain in a B-movie, complete with impassive face. I suspect he chooses his suit and tie carefully. When he's about to tell the world that he's having nuclear weapons made ready, for example, he opts for a black suit and black tie with tiny white dots, as if he's going to a funeral. Putin has been drenched in the ideas of Dzerzhinsky all his life and he's an absolute past master at those dark arts.

'At the Dzerzhinsky Academy you were taught the techniques of interrogation and torture,' Zsófia tells me. 'As a prisoner you had dark glasses or a blindfold put on you every time you were moved, even within the prison. You were kept in isolation at all times. Most important of all was the psychological terror. That's what everyone who's been tortured says: mental torture is far worse than physical torture.'

Putin is now applying these methods to all of us. He blindfolds us, so that we don't know what's coming. He plays with us. He's very good at that. With the announcement that 'something terrible will happen' as soon as NATO gets involved in the conflict, he feeds our fear. Putin is a dog; he can smell fear a long way off. You have to treat him like a dog that needs training: consistently, clearly, sternly, avoiding any gratuitous affront. This is one thing that makes the Ukrainian president and the Ukrainian people so extraordinary, and we should thank them on bended knee. Because let's not delude ourselves, without the bravery they've shown, we would all have rolled onto our backs in front of Putin long ago like a submissive bitch.

'Everything was focused on breaking people and destroying their dignity,' says Zsófia. 'The show trials and the interrogation and torture techniques were derived from those of the Soviet Union. The methods applied in Hungary and Romania came out of that same instruction manual. Aside from a bit of local ingenuity, the torture methods in the satellite states were identical. Electric shocks were applied to the body, with a lower voltage used for the mouth and the nose. Fingernails were pulled out and bodies seared with gas burners, the kind used as camping stoves. Cigarettes were routinely stubbed out on a prisoner's skin, the lips being especially popular because of their sensitivity. Prisoners were hit with truncheons, often on the soles of their feet so that they could barely stand afterwards. A special cudgel with balls of lead in it was sometimes used. Another frequently deployed technique was to shove a glass ampule, a hollow pipet, into a man's urethra and smash it. That was agonizingly painful, but it left no trace of violence. Something similar was done to women.'

If you're trained to pull out nails and to insert glass ampules and smash them, then the bombing of civilian targets like Mariupol or the use of chemical weapons in Aleppo is a trivial matter. I fear Putin will use all the means at his disposal to get his way and intimidate his opponents until they're on their knees – from insincere negotiations intended to divide his enemies to the use of tactical nuclear weapons, and blackmail by means of an exceptional number of civilian deaths. Because that's how the FSB works: the end justifies the means.

Budapest, Tuesday morning, 15 March 2022, 10.00 hours

As plentiful as Russians

Today the revolution is being celebrated in Hungary. The revolution of 1848 that is, against the Habsburgs, which failed because the Russian tsar sent reinforcements of 200,000 men.

The war is close by, not just because Hungary borders on Ukraine but above all because of my days at the border, the fact that I have two Ukrainian families staying at my home, and the continual stream of news. Also because I'm in three Ukrainian chat groups, with hundreds of updates from all over the country. Since my first night in hotel 'Little Ukraine' on 25 February, the second day of the war, I've been on Telegram. An avalanche of images.

A ruined, still smoking village with a dazed black dog lurching between piles of rubble. A brown teddy bear in tatters. The yacht *Lady Anastasia*, property of a Russian oligarch, offered for sale for seven million euro. The bodies of lifeless Russian soldiers with lumps of raspberry-red blood on their foreheads and collars, several of which have been laid out in the snow in the form of a Z and photographed from above. Old women blocking Russian tanks. Instructions on how to make a Molotov cocktail (crumbled polystyrene, petrol or alcohol, possible acetone too, bottle and rag). Instructions on how to throw a Molotov cocktail and where the vulnerable places are on various military vehicles. A saint with an anti-tank weapon, St. Javelin.

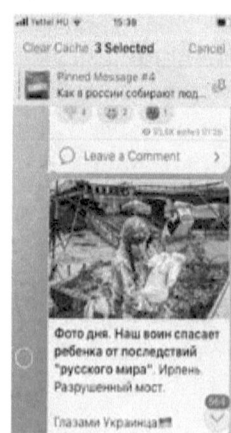

The chat groups are run by the Ukrainian ministry of defence, or allied to it. I don't speak a word of Russian or Ukrainian apart from *'Russkiy voenny korabl idi na huy!'* Very occasionally I translate a bit of text with copy-paste, but generally I just look at the films, at the photos and above all at the body language. You recognize the Russian soldiers immediately by the hopelessness they exude once caught, their bowed heads, the rags they generally walk about in and their youth, too: boys just out of school. The Ukrainian troops look eager to fight, mostly unshaven or bearded, strapping guys, older, in camouflage uniforms, well equipped. To judge by their body language they'll be able to keep up a guerrilla war for a long time yet.

Then there are the videos. Handcuffed soldiers being led away. The opening of Russian army food crates, zooming in on the use-by date: 2015. An armoured personnel carrier being hauled away by a tractor. Beautiful women fluttering their eyelashes as they sing to you about tank columns being blown up by Bayraktar drones. Night-time images of rockets flying horizontally over several blocks of flats before hitting one of them, a calm female voice in the background. A column of abandoned army trucks, a soldier showing their contents: shields, helmets and the long batons of the riot police. Protests at morning rush hour in Belarus, with hundreds of cars tooting their support. And since last night a lot of images and reports about the brave editor who burst into the Russian news studio with an anti-war message: Marina Ovsyannikova.

Today there's a series of photos of young soldiers going to the front, saying goodbye to their girlfriends. The situation is grimmer than ever, though. More pictures of collapsed buildings, fewer jokes, fewer naked traitors taped to lamp posts (there's no time for that now), fewer images of courageous civilians holding back Russian tanks. There are fewer pictures altogether. People are concentrating on survival, and they're probably also afraid of sharing too much information. It seems the Ukrainian ministry of defence is slowly getting something of a grip on the total anarchy shown in the avalanche of images of the first three weeks of the war.

My favourite films – it's terrible, I realize that, because they're just boys, sitting in those tanks; it shows how partisan I am, how much I see this as a battle between good and evil, how much I support the Ukrainians and am being drawn into their fight against an enemy that in terms of military materiel is twenty times stronger – show tanks being destroyed, either with anti-tank weapons hidden in the shrubbery or at night using Bayraktars. Sometimes there are films made by a small group of men who take you with them as they approach tanks and shoot at them. You find yourself right in the middle of the action.

They're winning, the Ukrainians, that's clear, but at what price? In the Russian army, morale is low and corruption reigns. But the disturbing thing is that Putin has endless supplies and doesn't care a jot about human lives. They'll keep coming, the columns, the rockets – and above all the bombing of cities will continue. There's a Hungarian saying, or actually a Szekler saying (the Szeklers are the most easterly of the Hungarian peoples, in what is now Romania, right up close to Moldova): '*Annyian vannak mint az oroszok!*' It's an expression used to say that something is present in abundance: 'They're as plentiful as Russians.'

Somogy, Tuesday 15 March 2022

Medicine for the people

Yeva's favourite government spokesperson, and very many Ukrainians feel the same way, is Oleksii Arestovych. He's an actor, columnist and blogger, but above all he's medicine for the people. Which is something the Ukrainians can do with.

In the Ukrainian chat groups on Telegram today, I see yet more images of destroyed Russian tanks but fewer bombed buildings. I suspect that Putin is holding back on the day before NATO and the EU come together, so that the gathered government leaders don't have images of bombed hospitals and wounded children too fresh in their minds. He'll no doubt carry on at full force again tonight and tomorrow.

Yeva and Alexey's children play outdoors endlessly, climbing onto and into anything it's humanly possible to climb onto or into and running around with our dogs, who think it's great. Mind you, the energy of three little Ukrainian boys – who were shut up in a VW van for days on end and then in our apartment in Budapest – is sometimes too much even for the young dogs. Alexey and Yeva sit on the guesthouse veranda for most of the day with their phones. The house doesn't normally have an internet connection, since it's used by friends and relatives who come to this remote village for nature, horse riding and hunting, but Ilona has managed to get mobile internet installed just in time.

During the day we leave each other in peace and keep in touch through WhatsApp, then in the evenings we have a drink together. That's the routine. The drinking takes place outdoors, on the long veranda. We light a fire. It's close to freezing, so cold that our wire-haired dachshund likes to lie underneath the brazier. The Ukrainian guests seem oblivious to the cold, another thing that will make the thousands of Syrian troops drummed up by Putin, in his desperation, despair of this war, the cold and the mud. Let's hope it carries on freezing at night for a while yet in Eastern Europe.

The other Ukrainian family, staying a short distance away with my parents-in-law, joins us. She, Yelena, greatly enjoys the howling of the jackals (it's like the howling of wolves, but at a slightly higher pitch). She laughs with delight as she mimics them. This is one of the darkest places in Europe, with minimal light pollution at night. The stars shine spectacularly above us. Alexey smokes his thin rollups; perhaps that's why we sit outside, as well as to avoid disturbing the children with our conversation. It's all made the oldest boy distraught and he keeps saying he doesn't want to die.

There on the veranda, Yeva shows me videos of Oleksii Arestovych. The Ukrainian government has about five spokespersons and I see them repeatedly in the chat groups. I can't understand a word they're saying, but the stalwart, unmoved impression they make inspires confidence in me nonetheless. They're unbeatable, and the Russians are still a long way from the government quarter. The Ukrainian leadership and the army are standing firm.

The speeches by Arestovych are far more important to Yeva and Yelena than those of Zelensky. Everyone who has crossed the border has left behind family, loved ones and friends in a country that's being bombed and violated, so they feel guilty, to a greater or lesser degree, about being safe. Arestovych speaks very softly, almost whispering, very calmly and, I'm told, in beautiful Ukrainian. His hushed tones are hugely reassuring.

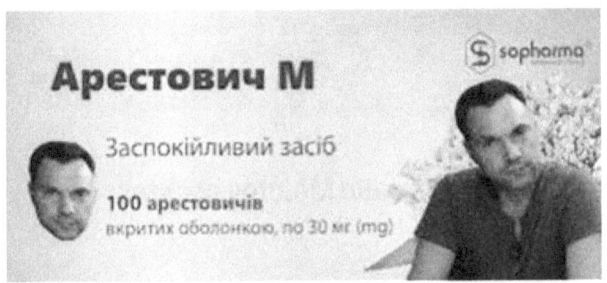

'It's like a sleeping pill. You sleep more peacefully afterwards,' Yeva explains to me. Arestovych usually sits slightly slumped in his chair, has short hair and wears a dark-green sweatshirt. A yellow-and-blue flag stands behind him. He's not a handsome man, but Yeva and Yelena's eyes sparkle when they imitate his voice. Yeva shows me that there are even photos circulating, on Telegram and in the Mamas chat group, of pill packets depicting his face – tranquilizers for 44 million Ukrainians.

Somogy, Wednesday 16 March 2022

How to make Molotov cocktails

Throughout these first few weeks I fall asleep and wake up with the war. I continually check the Telegram groups and what the BBC and *The Washington Post* (for which I've had a subscription ever since Catherine Belton, former Moscow correspondent for the *Financial Times*, started writing for it) have to say. Because of my recent search behaviour, I'm appearing on different radar. Until recently the internet mainly made me unsolicited offers for stair lifts and prostate operations, but since the war started all that has changed. Now when I type H and O into a search engine, some algorithm or other fills it in as 'How to make Molotov cocktails'. In the autumn of my life the world wide web thinks it can identify a sort of Che Guevara in me.

It's almost a month since I drove to the border. Hanging around there for four days and nights inspires humility. I saw people walking across with small bags of belongings, robbed of everything except their own skin. I saw teenage mothers with babies in rickety prams traipsing along the hard shoulder, swallowed up by the night. I saw Black students getting into buses with steamed-up windows. I saw mothers in SUVs with blow-dried hair and rows of children in the back. The men usually drove with them to the Ukrainian border and then got out to return to the war. Imagine the throat-tightening atmosphere in those cars, in the last few kilometres before the border, the parting at the roadside or in a parking spot.

Until that moment I'd manage to keep warfare a long distance from me for a lifetime. In my years as a schoolboy and student there was the threat of nuclear war, which inspired a stoical attitude, less than excessive ambition, and a love of Joe Strummer and The Clash. Later, after the Wall came down and war broke out in Yugoslavia, I fell in love in Budapest. The Serbs, Croats and Bosnians who had fled that war, whom you ran into in the nightlife of Budapest, were more than anything a confirmation that life is a great adventure. I read Hemingway and Orwell about the civil war in Spain. I don't remember much of *For Whom the Bell Tolls* except for some sneaking about in the night near a bridge, and of *Homage to Catalonia* only the British consul who refused to leave Málaga until the last possible moment, despite its merciless bombardment by Franco's army, because he believed the eyes of the world must be there.

That consul's story stuck with me, as an example of how to behave as a human being should you find yourself in such a situation and have the balls for it. Two Ukrainian journalists, Mstyslav Chernov and Evegeniy Maloletk, who arrived in Mariupol half an hour before the war started and stayed until the last possible moment to record for the world what happened there, have immense courage. In the end, at night, when time ran out – because Mstyslav and Evegeniy, as a result of their reporting in Mariupol, were at the top of the FSB list of people who must be tortured and killed – they were smuggled out of the besieged hospital by Ukrainian soldiers.

The Ukrainian punk rock band Beton has rewritten the number 'London Calling' by The Clash as 'Kyiv Calling'. The title of the original refers to the broadcasts by the BBC World Service during the Second World War that began 'This is London calling'. The Clash have given their approval for the cover version. The members of Beton are in Kyiv, and at night they help the wounded in the shelters. They know the raw reality of the war from close by, as reflected in the rewritten lyrics: 'Kyiv calling to the faraway towns. Now war is declared. A battle coming down. Kyiv calling to the whole world, come out of neutrality you boys and girls. The Iron Age is coming. The curtain's coming down.'

The war is growing increasingly dirty. The first evening that Alexey and Yeva invited us – they'd just arrived in our guesthouse, the war was a week old – that wasn't yet the case and the mood was cheerful. Ukraine wasn't yet lost. We sat at the table. Babushka, Yeva's mother, was busy in the kitchen and the smell of toast wafted in. Alexey had asked for an axe that day. He'd split wood and lit the open fire and the tiled stoves. The wood was burning and Alexey had bought a bottle of good wine from Villány. Babushka put toast with caviar down in front of us, Ukrainian caviar, one of the few things she'd brought from Kyiv in that tearing rush – a babushka knows what will come in useful.

First Alexey and Yeva told the story of the chicken, then they talked about the Roma from Kakhovka who were the first Ukrainians to succeed in stealing a Russian tank near Kherson, using a tractor, a method much copied after that. On the Telegram chat groups, I've since seen at least

eight different tractors dragging tanks and armoured personnel carriers across fields and through woods and villages. It's a cheering sight every time: a machine made for annihilation, for death and destruction, being towed away by a machine made for sowing, harvesting, life.

That was a turning point. Those were the days in which it became clear that Ukraine was going to hold out against a superior force, and that unarmed villagers were stopping tanks and Russian columns. Russian brutality had not yet been given full rein, with the bombing of hospitals and civilian targets. Suddenly there was optimism, a realization that the Russians were not invincible. It was a rare hope-giving interval between phase 1 of the war (aimed at military infrastructure) and phase 2 (aimed at civilian infrastructure). The fire, the wine, and above all the stories of civilian bravery and widespread solidarity made us euphoric.

Yeva launched into the heartening saga of a tough old Ukrainian peasant whose house was raided by Russian soldiers. She made tea for them and, good patriot that she was, didn't neglect to add a generous dash of laxative. When a Russian was sitting on the latrine behind the house, she poured a trail of petrol, splashed it against the wooden dunny, set light to it, tossed the jerry can inside too and burned the Russian soldier alive on the privy. These brand-new myths inspire the Ukrainians and make the Russian troops even more irresolute than they already were. As a Russian soldier, do you dare to put a cup of tea or a glass of apple cider, offered to you by a friendly, smiling farmer, to your lips?

Whether all these stories are true isn't really relevant. What matters is the invincible morale of the Ukrainians. I see it as completely intact, even now, in week four, despite the brutal performance of the Russians. What we in the Netherlands get to see on television and in the newspapers is a very different war. We see the victims and the refugees, while the Ukrainians see the heroes and the unexpected support from far flung places. We like to show people suffering, that's our national character, a preference for the pitiful. Here in Eastern Europe they prefer to show the victors. On Telegram I've been watching a war full of glory and deeds of heroism for a month now.

On the BBC I heard the story of an elderly woman who brought down a Russian drone worth three million euro by throwing a three-kilo preserving jar full of gherkins at it. Later the story was rectified. The glass jar had contained not gherkins but pickled tomatoes.

Somogy, Saturday 19 March 2022

Hair dryer

Alexey and Yeva and the other Ukrainians have left us. The opportunity arose to settle elsewhere and they took it. The day after they left, I was in Budapest, and in the Budapest-Ukraine WhatsApp group I made it known that we now had space again for one or two families. I received a message that at six o'clock that day two mothers, both called Yulia, each with one child, would arrive on the train from Záhony. They were looking for a place to stay for a week. I consulted Ilona; she was out there in the countryside and would make sure the guesthouse was ready and buy in some food. I let it be known that I could fetch the two families and accommodate them for a week.

In the hours that followed, their estimated time of arrival kept being adjusted. Eventually the train from Záhony came into the Nyugati station in Budapest at midnight. I'd parked on Teréz Körút and fetched pizzas and caprese from a good Italian, because they were sure to be hungry after a day in the train from the border. It was still miserably cold. The wide Teréz Körút was empty as I crossed it by the light of the streetlamps. Nyugati is a large, beautiful nineteenth-century terminus where trains arrive from the east, west and north. I was early, and just to be sure, I bought four bottles of water and put them in the car.

The long string of carriages came into the station and an army of the unkempt poured out holding sports bags, wheeled suitcases and bundles tied with string. By SMS I'd had erratic contact every couple of hours with the two Yulias. Walking against the human current, I went in search of two women and two children. The SMS contact had broken off. The rail passengers were directed by the police towards a side entrance in the direction of Váci út, through what was once Empress Sisi's waiting room. A broad, high passageway led to a cross-shaped hall with a cupola; on both sides stood aid workers handing out water, chocolate, nappies and other essentials. It was packed. I felt as if I was in a film about the Second World War, searching over the heads for loved ones before losing them for ever. Outside the doorway stood a cordon of police attempting to keep the indigents of Budapest away from the aid workers' stalls. The Puczi Béla tér was full of people looking on in a daze.

    I fought my way back to the station concourse and there, on an almost empty platform, I found the two bushwacked Yulias with their children, Bogdan and Sasha. In the car they were pleased most of all with the water; there had been nothing to drink in the train. Aside from 'thank you' they spoke hardly any English. I drove out of the city, across the Elisabeth Bridge and southwards to the M7, and asked whether their husbands were at the front. No, they didn't have husbands. Aside from Alexey and Yeva and the other couple that stayed with us, I haven't come upon any Ukrainians who are still married or whose parents are. Having children and then each going your own way seems to be the norm there.

    Fairly quickly the whole car was asleep and I was pretty tired too, so I drove fast. At Kaposvár, at one or two in the morning, five deer stood on the road just after a bend, stock still, frozen in the headlights, in a line one behind the other. I braked but could only go straight on and crash into them. Everyone was startled awake by the braking and the impact. I parked at the roadside, left the engine running, got a torch and checked the front of the car. On these B-roads through the forest you need to be on the lookout. The countryside is teeming with wild animals around here. I always try to watch for shining eyes in the verge, but if the animals suddenly appear in front of you, you don't stand a chance.

The radiator was leaking badly. I would need to hurry, so I ran ahead, looking for the deer. I didn't want to leave it lying in the middle of the road. It might still be alive, although that seemed unlikely given the force of the impact. It was fifty metres back, stone dead but without any visible damage. I dragged it through the bushes beside the asphalt by its front legs and laid it down carefully in the undergrowth. Then, as quickly as possible, I got in and drove home at sixty kilometres an hour, before the engine started to cook. I pointed the way to the guesthouse.

The next morning I checked the car. A big black stain had appeared under the engine. Strips of plastic and metal were bent, the radiator was dented, and at the bottom of the radiator were grey-brown deer hairs, as if swept together on a hairdresser's floor. Later that day the car would be fetched and then be gone for almost a month because nobody anywhere in Europe had a radiator in stock. I was pissed off by the cost and by being without the car for so long.

I walked over to the guesthouse to see if the two families needed anything. In the village we have a shop with an old-fashioned socialist-austere decor where you can buy basic supplies like flour, washing up liquid, toilet paper, long-life milk, cans of beer and white bread. For exotic items like muesli, bananas and avocados you need to go to Kaposvár, thirty kilometres away. The two children were happy with the visiting dogs, and the joy was mutual. The dogs were missing Misha, Gosha and Jenya.

Yulia stuck her index finger in the air; I mustn't leave yet, there was something important. She got out her phone and typed a message, feeding it into Google Translate. She looked very serious and took a long time about it. I was expecting a request of an existential nature – as indeed it was. She straightened her arm and stuck the phone under my nose, so that I could read the message: 'Good morning, tell me please, where can I buy a hair dryer?'

Somogy, Sunday afternoon, 20 March 2022, 16.00 hours

List

The winter wheat brings the bright-green glow of spring. I understand that in Kyiv the market is open as usual and people are buying mainly seed. They believe in the future. Despite the claim of the former communists that they took such good care of their citizens, Eastern Europeans have a deeply rooted distrust of governments and an accompanying awareness of the need for self-reliance. It's a feeling we have lost in Western Europe. We regard it as a matter of course that the government will pamper us from cradle to grave. Here, in the former Eastern Bloc, there are no such high expectations.

In Hungary, just as in Ukraine, we are sowing. Almost everyone has their own vegetable plot with salad greens, tomatoes, sweet peppers, onions and potatoes, along with fruit trees. Or if they don't, then their parents or grandparents do. In summer people make jam from plums and peaches, and preserves for the winter from peppers and tomatoes. In our village it doesn't make much difference if the electricity supply is interrupted and cash machines cease to work. Television, the Champions League, Formula 1, video games and pornography will all be missed, but life goes on. Every adult has a legal right to produce twenty litres of home-made palinka each year, which helps to pass the time. Money is not essential; the village barter economy works, so a few chickens can be exchanged for a jerrycan of diesel to keep the generator going. In every garden there are piles of black locust wood to heat the house.

In the first four days of the war I saw the overwhelming help for the refugees coming across the Ukrainian border given by Hungarian and international NGOs, and by individuals, and concluded it would be good to try to support the victims and fighters in Kyiv. In Budapest, along with a group of friends, I started organizing accommodation for refugees, and I got to know a Brit who had fled Ukraine. He put me in touch with Dima and with a medical student in Kyiv.

I received a message from Dima in Kyiv, saying, 'It's relatively quiet here, apart from artillery noise; you can hear it at any time of the day. Good thing it's Ukrainian military. Let me get back to you in a couple of days.'

A few days later I received the shopping list.

We need: QuikClot Combat Gauze, Israeli bandages, Trauma shears, HALO, Decompression needles, Military first aid kits, Combat application tourniquets – PRIORITY, Celox.

As well as other combat gear: Helmets, Goggles, Knee and elbow pads, Combat gloves, Weapon cleaning kits, Rangefinders, Sights for calibre 14.5 (7 archer), Quadrocopters, NIJ3A armour and Webtex vests.

Budapest, Thursday evening, 24 March 2022, 19.00 hours

Day of Judgement

The shopping list from the front included among other things bulletproof vests, night sights and quadrocopter drones. At first I tried getting hold of some of the listed items online, but it soon became clear that I couldn't expect my first delivery of trauma kits and knee pads from Amazon until the start of May at the earliest. A bit late when you're trying to save lives. By that time all of Ukraine might have been bombed to pieces. My online search then spontaneously turned up Turkish and Israeli dealers who were offering bulletproof vests for six hundred euro. With my ingrained suspicious streak, partly a result of having grown up in the Dutch province of Twente, I had little faith in that.

Having asked friends to help me with money and contacts, I was hopeful of finding the much-needed protective equipment quickly and went in search of a place where I could store the supplies before taking them to the border or to Poland. Inner-city Budapest is an awkward place for deliveries – narrow streets with few places to park – so I decided to visit our old cook. She lives on the edge of the city with a lot of space around her house. As ever I'm invited to sit in the deep sofa with a view of the wall with family photographs. First a cup of coffee. She's a treasure; she always wants to spoil you. The milk for the cappuccino has to be carefully frothed, while I chat about incidental matters with her husband. She's the boss, the one I need to consult.

They say it's okay if things for Ukraine are delivered to them. I've stressed that they'll be medicines and I keep quiet about the other things I'm working on for the time being, because they're Jehovah's Witnesses and opposed to weapons and war. I'm not going to be delivering weapons, but there will be helmets, bulletproof vests and night sights – it's a fine distinction. Years ago, when they both worked for us, it was only with gentle pressure that I persuaded the cook's husband to fix gold epaulettes to his butler's jacket – they made the uniform too military – so it's better to keep quiet for now about the kevlar helmets we're trying to buy from the Italian army. In any case, I'm still in the process of ordering them; it's all a good deal more demanding than I thought it would be.

Since I live not far from the Ukrainian border and have studied the diabolical methods of Soviet repression, I feel more than most a sense of urgency when it comes to supporting the Ukrainians to the maximum. I feel it more than the majority of Hungarians, who underwent that repression not long ago and now prefer to stick to peaceful talk. But Putin must be stopped. And you can do that only with concerted, tough action. He's like a pit bull, the sort that respects you only if you give him a firm slap.

As I sink deep into the sofa with my cappuccino, we talk about the war. These people, for whom I have great affection, our cook and our gardener, who cared for our children for almost fifteen years as lovingly as grandparents, stick up for Putin. And not just a little bit either. It's the fault of the Americans and the Ukrainians, this war. Putin doesn't mean very much harm. It's as if I'm listening to a commentator on Russia Today.

I'm shocked. The bombing of Mariupol is entirely the fault of Zelensky. I try to contradict them, but my Hungarian is really only good enough for a casual chat about this and that, about nature, the weather, food and structural alterations, not geopolitics. Fifteen minutes later my stomach hurts as I drive down the hill, the entrancing city with its thousands of lights laid out before me and above it a purple sky like a watercolour by János Vaszary. How is this possible? Is their opinion attributable to the pernicious influence of Hungarian state television, or has it all been thought up by the Jehovah's?

There's one thing I do understand. That if you spend your whole life waiting for the Day of Judgement, for the day when the wheat will be separated from the chaff, for the day when those decades of living strictly according to the rules of *The Watchtower* are at last rewarded, if you secretly can't wait for 'the end of the kingdoms of this world' – then you're likely to place all your bets on Vladimir Putin.

Somogy, Friday 25 March 2022

Lavender and shit

I recently went to Laci, the marble salesman a few villages away. He's from Transylvania, from Marosvásárhely (Târgu-Mureș in Romanian), and he has a slight lisp. I've been visiting him for twenty years, whenever I need hardstone. His main income is from the sale of gravestones and grave monuments. He imports marble from Italy and has old slabs of it stacked up, for kitchens and bathrooms that were never collected. From time to time, for a hundred euro, I fill the boot of the Land Rover with plates of Carrara marble that have been sitting at the back of his yard in all weathers. Sometimes they're a bit brittle, but they always have a patina.

Laci wanted to show me the lavender plantation he's laid out around his house. On the way there, along a dusty, winding gravel path between meadows and woods, we got talking about politics. He too turned out to be extremely anti-American, just like our old cook. Putin was not completely okay, but compared to the Americans he was a kind of angel.

'Everywhere the Americans go there's a war. The war in Ukraine is because of Zelensky, but above all because of the Americans!'

I decide to check out this worldview, so different from mine, by ringing three Hungarian friends and a Dutch friend who has lived in Hungary for many years. The first, from a family that votes socialist, is a lifelong opponent of Viktor Orbán's party, Fidesz, and he agrees that the Americans are most to blame for this war, with their promises of NATO membership and too big a presence in Ukraine.

'The Americans provoked this. It's a war between America and Russia. We don't have anything to do with it and mustn't interfere.'

The second friend is a Hungarian who grew up in the Netherlands, the son of two Hungarians who fled in '56. He hates both the socialists and Fidesz for their lies and corruption.

'The Hungarians are cowards. They lap up Orbán's talk about staying out of the war. He sets himself up as the great peacemaker, as the protector of the Hungarians, as a statesman who can keep Hungary out of the conflict. The war is presented as a fight between America and Russia, with the Eastern Europeans as its victims. Everyone is frightened of losing the luxurious life they've built for themselves since the fall of the Wall. Orbán's going to win the election by a landslide as the preserver of peace for Hungary. I'm afraid he'll get a two-thirds majority. Then we'll be the losers.'

The Dutch friend, a devout Catholic, married to a Hungarian, wastes no words.

'I think a serious investigative journalist ought to scrutinize how Hungary makes its payments for Russian oil and gas. No one knows exactly what the arrangements are. It goes through dealers in Switzerland. They buy it from Russia and the Hungarian state buys it in turn from those dealers. Nobody knows what the price is and who is involved. Creaming off European subsidies might well be child's play compared to the sums earned in that process.'

Lastly I speak to Aron, a friend whose judgement I greatly respect. He's studied psychology. He likes neither Fidesz nor the socialists and hates the craving for power, the corruption and lies of both groups, like any right-thinking person. But there's no sign in him of the foaming rage that reveals itself in so many people, both in Hunga-

ry and in Western Europe, as soon as the subject of Hungarian politics is raised. Few people are capable of looking at it with a degree of detachment; everybody immediately has a pronounced opinion ready, copied from the media. Aron is the exception.

The Hungarians blame the war on the United States, Aron says, because the Hungarian newspapers, radio and television keep repeating that message. Another oft-repeated tune is that Hungary needs to be particularly cautious in its response because otherwise the 150,000 Hungarians in the Ukrainian province of Transcarpathia will be in danger. They might suffer Russian retaliation. This theory holds water only if Putin conquers the whole of Ukraine, including the most westerly part. In reality the Hungarian minority is now the victim of Orbán's politics, because the Hungarians, including the minority in Transcarpathia, are regarded by the Ukrainians as lacking solidarity with the Ukrainian cause.

My own fear is that if Putin wins the war – which I don't think he will – he'll give Transcarpathia, which was part of Hungary until the Treaty of Trianon, to the Hungarians as a gift. Putin isn't interested in the 150,000 Hungarians there, or the tomatoes and potatoes in that westernmost part of Ukraine; there's no industry there, and no mineral resources. It's the ideal way to split the Europeans and it won't cost him anything. I doubt whether the Hungarian leadership could resist such a temptation. It would mean the end of the European Union in its current form.

'Aron, what do you think? Are there many Hungarians who'd want Transcarpathia to be returned to Hungary?'

'Ach,' Aron answers. 'I don't think there are many Hungarians who seriously want Transcarpathia back or believe that it will be returned to Hungary, thank god. And I doubt Orbán wants it to be; he's not that stupid. But he's got a big mouth, that's his problem. There's a well-known Hungarian fable. A sparrow gets stuck in a cowpat. The sparrow makes a huge scene, a great racket. Then a cat comes to investigate the noise and eats the sparrow. It's a parable about Hungary. With its total dependence on Russian oil and gas, Hungary is in deep trouble. The lesson to be drawn from the fable is that when you're in deep shit it's best not to shout too loudly.'

Bratislava (Slovakia), Saturday afternoon 26 March 2022, 18.00 hours

Between East and West

I've suspected for quite some time that Viktor Orbán has taken Gábor Bethlen (1580-1629) as his role model. Bethlen is the Transylvanian prince who inspires him, a ruler who brought Transylvania wealth and fame, and was a master at balancing between East and West, between Constantinople and Vienna.

Soon it will be five hundred years since the centuries-old Kingdom of Hungary lost its independence to Sultan Suleiman the Magnificent at the Battle of Mohács (1526). Central Hungary, including Budapest, was occupied by the Ottomans. The north of Hungary – including what is now Slovakia – retained a kind of freedom, under Habsburg protection, with Pozsony, today's Bratislava, as the temporary capital of the amputated kingdom.

I'm in Bratislava for the afternoon, and most of the impressive old buildings you see here are Hungarian, built by magnates who had names still used by people we come upon from time to time at parties in Budapest: Zichy, Esterházy, Pálffy, Desewffy. Fragile elderly gentlemen from a vanished world. The houses built by their forefathers, palaces painted pale-yellow and white, are now in use as Slovak government buildings, surrounded by grounds with high walls featuring beautiful stone carvings of vases every fifteen metres.

After defeat at the Battle of Mohács, mountainous Transylvania became a semi-independent princedom un-

der the sultan in Constantinople. It had to pay taxes to the Sublime Porte and the sultan had a veto on the appointment of the prince of Transylvania, who was chosen by the nobles from among their number. If the prince or the nobles did something the sultan didn't like, he would send the Crimean Tatars to Transylvania on a punishment expedition, to teach the Transylvanians a lesson by murdering, raping and looting. I imagine this is roughly the model that Putin had in mind for Ukraine on the eve of the invasion: a leader of his choice, taxation (or otherwise a share of the income from gas, coal, the steel industry and agriculture) and instructive punishment expeditions from the Crimea should Ukraine deviate from the behaviour required of it by Moscow.

As a semi-independent princedom from 1570 to 1711, Transylvania had to navigate a middle way between the Ottomans and the Habsburgs, because they too were trying to exert their influence and the two empires could be played off against one another by the prince. The Transylvanian nobles were mostly Protestants, with approximately the same motto as was found right across Europe in those days, even shouted from the walls of besieged cities in Holland: 'Rather Turkish than Papist.'

The princes of Transylvania tried to pay as little tax as possible, and to acquire as much freedom as they could without having the pillaging Crimean Tatars unleashed on them. Gábor Bethlen was a past master at it. He was an enlightened despot, and he presided over a marvellous period of affluence and festivities. He developed mines and industry in Transylvania, and acted as a patron of the arts and protector of the Calvinist Church. Bethlen nationalized much of the country's foreign trade and set up a network of agents, who purchased products at a fixed price and sold them abroad at a profit.

In Western Europe people love to portray Viktor Orbán as evil personified, forgetting that he's brought Hungary and some of the Hungarians a number of good things: stability and economic growth, a reduction in the huge national debt, the restoration of a kind of pride in being Hungarian (just as Putin did for the Russians), a settlement of the personal debts that hundreds of thousands of poorly informed Hungarians had run up in foreign currency, the rule that a certain percentage of items sold in the supermarkets must be Hungarian in origin, and finally, controversially of course, the low price of oil and gas from Russia.

Orbán is alone in his family in being a member of the Calvinist Church; the rest are Catholics. How deeply rooted his religious feelings lie is impossible to judge, but the fact that he has converted to the same denomination as Gábor Bethlen supports my hypothesis. There are many other ways in which he mimics the prince of Transylvania, as a new enlightened despot. Orbán too is a shrewd businessman who has set himself up as a patron of the arts, which is to say of the arts and artists that meet with his approval, and as a patron of soccer and everything that is Hungarian. But it seems he's not a patch on Gábor Bethlen when it comes to balancing between East and West, to playing them off against each other and thereby getting the most out of the situation both for Hungary and for himself.

Since coming to live in Central Europe, I've learned that geopolitics is ultimately the deciding factor. Before Romania and Bulgaria joined the European Union, Hungary was the West's most easterly outpost, the model pupil, and the place that attracted all the foreign companies and investments. Hungary has lost that position since 2006. It's become relatively insignificant now that Romania, Poland and the Baltic states are of far greater importance in geopolitical terms. From that perspective, the way Orbán has continually focused the world's attention on Hungary represents quite an achievement.

For more than a decade, Hungary has sought a rapprochement with Asia. The Asiatic brotherhood and the fact that Hungarians originated in the Asian steppe have been cherished and celebrated in state visits and by the Kurultáj, an annual meeting of Central Asian cultures and peoples. Horse riding, archery, flag flying, handlebar moustaches and headgear featuring a lot of fur are all encouraged and appreciated. In parks, in playgrounds and at commemorations you see a heroic past recalled, when tough guys on horses galloped across the steppe.

Aside from such folkloric festivities, there are other things that point the way: close ties with Putin (which are probably mainly pragmatic), and the moving of the headquarters of the Russian-led International Investment Bank from Moscow to Budapest. The staff of that bank, which is regarded as a huge base for the Russian secret services (it's the first time that such a Russia-dominated institution has set up its HQ in a NATO country), enjoy diplomatic immunity. There's also the major Russian involvement in the Paks nuclear power plant, Hungary's complete dependence on Russian gas and oil, and finally the planned founding in Budapest of a campus of Shanghai's Fudan University for six thousand students – a new centre for research and intellectual property theft in the heart of Europe. All these choices taken together do not feel like innocent flirting but more like leaping into bed with someone, or at least stealing a kiss in the cloakroom.

It will be interesting to see what Orbán does over the coming year. Will he slowly turn away from Putin, or will he strengthen his ties with the East? It will be hard for Viktor Orbán to carry on playing at being Gábor Bethlen.

Medyka (Poland), Sunday evening, 27 March 2022, 20.00 hours

Exodus

I make my way through Slovak pine and birch woods in a hired car. It's as if I'm not in Europe at all but driving across Montana, with endless unspoilt nature all around. Slovakia is the perfect setting for a Quentin Tarantino film, with insane sculptures, empty shops and blocks of flats in remote villages, and communist nostalgia, including tanks on concrete plinths in memory of Second World War battles. The natural landscape doesn't change at the Polish border, but at last I see nuns on the streets again, with black veils. And the houses are different: broad, low structures and delightful old log cabins. This area was incorporated into the Habsburg empire late and only briefly, otherwise the wooden houses would not have been permitted. In the eighteenth century, Empress Maria Theresa promulgated a law which said that in Hungary at least (and therefore in Slovakia, which was part of Hungary at the time), houses must be built of stone or loam.

At the end of the afternoon I arrive at the Polish-Ukrainian border post of Medyka. It's a huge tented encampment, with city buses and food trucks. There are camera teams and tough-looking Polish soldiers in camouflage uniforms conducting the traffic and helping refugees into the buses. They are going to two places, an empty school in Przemyśl, where there are places to sleep, and an abandoned Tesco's supermarket with two thousand beds. Reception at the border is magnificently organized: a long street with a bend in it, and on either side aid agency tents, their stalls piled high with supplies.

It's a tent village, and on the receiving side it has the excited, fraternizing atmosphere of a music festival, but without the music. A vast warmed tent with the words 'Taking down the Chinese Communist Party is the only way to save humanity' written on the side is the first thing I see. The United Sikhs, with the slogan 'Recognize the Human Race as One', have wonderful long beards and turbans, and they fry excellent curly chips (which towards midnight will serve as my dinner). The World Central Kitchen hands out thick soup, oranges and Easter eggs. The Jewish Agency for Israel has the stand closest to the border crossing. Cadena has a stall with coloured pencils and children's toys. At the aid organization SEWA International, Hindustani in origin, dozens of phones can be charged simultaneously, and the Polish Red Cross is bursting at the seams with medicines, toilet paper and hand sanitizer. There is a separate tent for the reception of pets. Around me are tall stacks of sanitary towels, bottles of water, or cans of milk powder, along with mountains of old clothes. Cubicles containing chemical toilets have been installed here and there, the hum of generators sounds all around, a rotund man from American Acts of Mercy manoeuvres a bulky quad bike through the crowd, and an exhausted mother leans on the pram she's pushing westwards as the sun turns the sky pink as candyfloss.

It's overwhelming, even for me. How must it be for people from remote villages in the province of Zaporizhzhia, people who have escaped bombardment, crept out of shelters, fled for their lives, who suddenly find themselves in this colourful bazaar where everything is free and where, at least in the hours I spend there, the aid organizations compete with each other to shower the refugees with their empathy and gifts. Actually, if I'm honest, I'm not sure what I'm doing here. At the border between Hungary and Ukraine I'd come to pick up a family. I'm here because I needed to be in Rzeszów, which isn't far away, and wanted to see whether this place was very different from the situation on the Hungarian border. After walking twice along the aid workers' street I've seen it all. I feel like a disaster tourist.

I'm on the point of carrying on to Rzeszów when someone taps me on the shoulder from behind and asks whether I'll keep watch tonight. It's a tall, unshaven young man with a ponytail. Around him hangs the smell of a person who has been going about in the same clothes for quite some time.

'How d'you mean?' I ask.

'The gypsies. They come between one and seven in the morning and strip the camp bare. It's really outrageous, a swarm, every single night.'

He takes me with him to Mission Ukraine, housed in an army tent that's filled with the smell of sweat and palinka. Bullet-proof vests hang against the tent wall. Within five minutes he's dredged up his past, American style. The likeable hippie got on a plane in New York in the second week of the war without a plan, intending to help refugees, and after three weeks of voluntary work he's become integrated to the point that he talks as forthrightly about 'gypsies' as the average Eastern European. I tell him I'm just passing through and unfortunately can't serve as a watchman to prevent the Roma from making off with the bars of chocolate and packs of sanitary towels.

A month ago, after four days at the Hungary-Ukraine border, I came to the conclusion that it might be better not to focus purely on addressing the symptoms (caring for refugees) but on tackling the problem itself (the invasion of a country). I decided to help people inside Ukraine, and the army especially. In early March I received a list of things people needed as a result of the fighting and bombing. I had to get to grips with new jargon. A tourniquet is a device that can be screwed tight, as it were, to stop serious bleeding, whereas an Israeli bandage, invented by an Israeli army doctor, has a pressure applicator and a closure bar that enable it to continue staunching the bleeding by itself once applied. One of my brothers is part-owner of a pharmaceuticals factory, another is a doctor, a friend is a doctor in teaching hospital and another friend is an important physician in the Netherlands. I sent them the list from Kyiv and asked whether they could get hold of some of the items needed through their contacts, the directors of their hospital, or their hospital's back door.

We can only hope war doesn't break out in the Netherlands, because it seems life-saving battlefield medical supplies don't exist there. At least, none of the four managed to get their hands on anything at all, not even a bandage. Perhaps there will be a first aid kit with tourniquets and QuikClot Combat Gauze in some of the eighteen tanks the Dutch rent from the German army if the war comes our way. After a series of disappointing responses I concluded I'd have to purchase supplies, so I started collecting money.

Putin's ambition to turn Ukraine into a failed state will have succeeded as soon as he manages to chase a million articulate, highly educated people and other troublemakers out of the country for good – as the Russians did in Hungary in 1956, when they kept the border porous for several weeks, so that almost 300,000 Hungarians fled. That's what makes the consistent and targeted bombing of residential areas, apartment buildings, hospitals and schools by the Russians so worrying, quite apart from the criminality, the horror and the unfathomable suffering. It's intended to bring about an exodus, to burden Europe and keep it busy with a derivative problem.

If twenty million or more Ukrainians enter the EU, it's a minor disaster for Europe, because of the social unrest the influx will inevitably create, and a disaster a thousand times greater for Ukraine, because it's always the best – the people most prepared to take risks, the best educated, the richest, the youngest, the people who speak several languages, the rebellious and freethinking, the intellectuals, artists, doctors, engineers, computer programmers, the people you need to build a democracy – who ultimately leave their country to start a new life elsewhere.

The refugees leaving Ukraine and the essential goods, from medicines, ammunition, food and fuel to weapons, that are coming into the country and have allowed the Ukrainian armed forces to fight on up to now, move along roughly the same roads, in opposite directions. The fact that Putin has not yet bombed those roads, railways, bridges and viaducts sufficiently to block the flow of aid supplies shows how crucial the exodus into the EU is for him. Or how disorganized the Russian army is. One of the two.

I'm blessed. I grew up in an environment where there hadn't been a war since before I was born. At Medyka I see what war does. It's eight in the evening. People are crammed together like animals, in cheap skiing jackets, empty looks in their eyes. They are different from the refugees in their own cars that I saw in those first few days at the Hungarian border – poorer and more beaten down. It's the difference between several days and several weeks of war.

Medyka, Sunday 27 March 2022

A glorious death

It's dark now at the Medyka border crossing. Andrea, a pretty young Mexican-Jewish woman who works for a Mexican-Jewish crisis organization, tells me that here they call it the 'United Nations Festival'. There is indeed a cheerful, bonding atmosphere. Like me, she's incensed by the reactive attitude of the West and the paucity of serious weapons being supplied to the Ukrainian military. It's perverse; the West is offering them just enough for a glorious death.

Andrea also explains to me that, if I want to, I can enter Ukraine with one of the aid workers. She regularly crosses the border to make the wait in the freezing cold more bearable for people in the long lines on the Ukrainian side, by handing out blankets, soup, coffee and whatever else she has. The aid workers don't have to go to the end of the queue and they have a separate path for coming back into Poland. I join a group of Brits. I'm handed an army rucksack to take with me, help to lift a huge pan of soup into a shopping trolley, and then lose the Brits because I'm looking for my phone, which I've connected to a charger somewhere in one of the tents.

Half an hour later I pass two Polish soldiers in camouflage uniforms, who are standing at the gate where the refugee reception camp ends and the customs checks begin. My Uncle Chuck learned from Anthony Fokker that when you're poking about on your competitor's terrain, the most important trick is to behave as if you belong. I've got the

army rucksack on my back and I'm wearing a yellow high-viz vest – it's still got the folds in it; I've just pulled it out of its plastic packaging; I found it in the rental car that I drove to the border.

When the sun went down two hours ago, the sky turned old rose like my grandmother's tea service. Now the sky above me is deep cobalt blue. I walk along a broad path of hard brick between grass-green fences. The fence to my left is high, like the fence around a basketball court or a prison. The one to my right is two metres tall; if you climb over it you're on the vast stretches of asphalt of the customs area, lit up in bright white by a superfluity of lamps. There's a line of trucks and buses trying to get into Poland that still need to be checked.

Among them is an SUV with a horsebox. Might someone be trying to take a horse out of the country with them? It's possible. Horses are unaffordable now because of Covid. As well as Covid dogs, a huge number of people have got themselves a Covid horse. Cart horses with big haunches and crooked teeth that no one would previously have bought in the plains of the Carpathians now cost as much in the EU as jaunty crossbreeds before the pandemic. In the Medyka refugee camp there's a separate reception for animals, with a big tent behind it to shelter them. It's quiet there. The people fleeing now are from the countryside and the provincial towns, and they have a more practical attitude to their livestock, less sentimental than the city flat-dwellers who haul baskets of cats and dogs with them.

The path leads upwards. It's quite a walk, around five hundred metres. On the other side of the high fence a cluster of refugees walks down from time to time, on unsteady legs. Exhausted, dazed. There on the other side of the fence it's dark. There too, people walk between two fences. The outside fence on that side has a roll of wire on top, not barbed wire for keeping cattle but that vicious NATO razor wire. I walk calmly uphill, entirely alone. At this hour of the night nobody goes into Ukraine on foot. It feels like a rite of passage. I have time to contemplate my sins. It's a long upward path. With the army rucksack, the high-viz, the green hunting jacket under it and my sneakers, I look like a man intending to join the Foreign Legion. A romantic, an adventurer. Spectacles on my nose and autumn in my soul.

There on the other side of the border is the part of the world where Isaac Babel hunted Polish squires with his Cossack regiment. A hundred kilometres to the east is Lviv and further on Brody and Novohrad-Volynskyi. I walk towards the light, towards the little building that houses the Ukrainian border patrol. I go through a door and then through a steel turnstile that ensures you keep going in one direction only. A man who looks like a tramp tries to worm his way through from the other side. In vain. He's called to order by a soldier, pleasantly, as if he's a familiar local character.

I'm in a bare room, battered like a football changing room. In front of me is a corridor with two hatches manned by soldiers in camouflage. Once I'm past those, I'm in a country at war. To my right a man wearing a bullet-proof vest is standing in a corner with a shopping trolley containing two rucksacks, one black and one in camouflage colours.

This must be one of the men who have responded to the call by President Zelensky to come to Ukraine to fight the Russians. I hesitate for a moment, but then speak to him. He is Neal, a US Marine from Alabama.

'The Heart of Dixie,' he says proudly.

He left a wife and three children behind in Alabama and flew across, caught the train and walked here from the station. He sees it as a duty to stand up to the evil that Putin is doing in Ukraine. He's one of those men who won't be able to do much more than die a glorious death if we don't make sure the Ukrainian army and the Ukrainian Foreign Legion get the weapons they need to protect civilians and go on the offensive. We exchange telephone numbers and promise to stay in touch.

I'm on the Ukrainian side of the border. A horde of people stands pressed together in the darkness. Close to them are tables attached to the field kitchens set up by the aid organizations. I notice that hardly anyone steps out of the line to fetch any of the goodies on offer. Nobody wants

to lose their place in the queue for the border. Or perhaps they've become suspicious of the distribution of supplies. The Russians are sending trucks of food to conquered cities like Kherson, where it's distributed with a great song and dance, but the local Ukrainian population refuses to accept it. Russians or separatists from Donetsk have to be brought in by bus to make videos for the Russian evening news, showing how pleased they are by the arrival of their liberators.

I find my British aid workers again. They've got a shopping trolley full of bags of sweets and chocolate surprise eggs. I take an armful and walk into the line. Some people say a quiet 'thank you', but most don't speak a word of English. The queue is made up almost entirely of elderly people, women, children and the occasional teenage boy. I ask them where they're from. The provinces of Dnipro, Zaporizhzhia, Zhytomyr. There's nobody from Kyiv. The big city people fled first; now residents of towns and villages in the rest of the country are following.

It's difficult to get them to take the bags of sweets. They contain something like wine gums. I'm experiencing how hard it can be to act as a kind of patron. Many people don't want them; the odd child takes one from me gratefully. I feel absurd and superficial, even a kind of imposter, with my arms full of sweets and surprise eggs. One person goes to Ukraine to fight the Russians in the woods and the trenches, another to share out little chocolate treats.

Medyka, Monday morning, 28 March 2022, 01.30 hours

Father

'I fed him morsel by morsel with a spoon,' Michael tells me. 'He was an old man in a wheelchair, disabled, left at the border by his family the way you might tie a dog or a cat to the crash barrier. He couldn't do anything for himself. He had nobody at all, abandoned by everyone. When Lviv was bombed he suddenly found himself here, in a great flood of refugees.'

Michael has been at the border for three weeks and he concerns himself with elderly people, the lonely. The war brings good things too, such as Michael, who helps where he can, and who as a result of coming here has met his father for the first time. Michael is a Pole who is living and studying in the Netherlands, speaks fluent Dutch, English, Russian and Polish, and has returned to his native country to help refugees; a tall, pleasant, open young man, slightly high and hyper from all the excitement, like several of the aid workers in Medyka. I suspect he falls for men – he has a sensitivity that many heterosexuals lack. Maybe that's why he prefers to study in the Netherlands rather than Poland.

I walk back with him from Ukraine into Poland. He points the way. He's done this before. The people around us look exhausted. In the tight lines of refugees is a young woman with thick calves who is carrying nothing except for two violin cases, and an old woman with only a plastic bag in her hand. In the brightly lit customs office we squeeze past the refugees, who have spent hours standing in the freezing cold. It feels a bit like visiting a hospital in Hungary; as a Dutch person you almost always get preferential treatment, based on your name and appearance, and the doctor sees you more quickly than other patients. On the one hand the VIP treatment is of course nice (I don't exactly enjoy waiting), but on the other hand it's embarrassing to be beckoned forward and walk to the door of the consulting room past everyone else, looking suitably grim. We push forward under the strip lights. Michael goes to a separate hatch for aid workers with EU passports. We get a stamp and our papers are only cursorily examined. People in high-viz come and go constantly. Leaving the packed queues of the desperate behind us, we walk between high fences into the Polish night.

To warm up we drink a cup of coffee in the aid workers' village, in the heated tent of the 'New Federal State of China', financed by a rich anti-communist Chinese and an American. Michael tells me he too keeps trying to see beauty in the war. Michael's parents divorced when he was a baby and he grew up with his mother. He knew his father lived somewhere close to the Ukrainian border and ten days ago he rang him. Last week they arranged a rendezvous in a restaurant. Michael had never met him before. He looks at me, excited and joyful.

'It was great. After the meal we walked side by side along the street and lit cigarettes,' says Michael. 'I can no longer remember the first cigarette I smoked with my mother, but I'll never forget the first cigarette I smoked in the company of my father.'

Rzeszów (Poland), Monday afternoon, 28 March 2022, 17.00 hours

Nuns and kebab

There are nuns in the streets of Rzeszów too, and kebab stalls, lots of them. George Friedman, an American geopolitical strategist and forecaster who was born in Hungary, predicts that the Turks and the Poles will make a pact in the future. Are the kebab stalls a prelude to Turkish-Polish fraternity? When I arrived in Poland yesterday from Slovakia, through the Carpathians, it was as if I was driving into Transylvania: the same landscape, the same rolling hills, the same vegetation, but without the atmosphere produced by those many colourful stalls in the verges selling reed baskets, plastic garden gnomes and folkloric kitsch. In contrast to Romania, the Ottomans never got as far as Poland to colonize it. Perhaps that's why it lacks the fairground-style roadside bazaar.

Poland and Romania will probably come out of this war stronger. Both countries have pronounced anti-Russian sentiments and are proving themselves reliable allies of the United States. Turkey is an important NATO ally, certainly, but unpredictable and completely untrustworthy. I think that over the coming years the Americans will move as much materiel as possible, including nuclear weapons, out of Turkey to the Mihail Kogălniceanu air base in Constanța, Romania, to keep a check on the Black Sea and the Bosporus from there.

'A Ukrainian saying goes that if you want to make off with a lot of money, you should build a road,' Alexey, my new Ukrainian bosom friend, tells me. 'You can bury a lot of money under a new motorway. But a war is even better. Nobody knows where all the money is going.'

That's why I'm back in Poland, to take care that the protective equipment for Ukraine is handed over properly. I'm responsible for it; I've collected money from friends, from my former student fraternity, from my club, from my old rugby team in Delft and from a whole spread of acquaintances – the rugby players turned out to be the most generous by far. I and a friend in the United States have used the money to buy 125 kevlar helmets from the Italian army, 340 knee and elbow pads and 340 trauma kits including tourniquets. I feel like a Levantine Greek, an intermediary between the West and the Near East. In Constantinople it was they who brought together the two worlds: the world of signed contracts and the world of handshakes.

In Rzeszów I have an appointment with Ukrainians Illya, Symon and Lesya, of The Ukrainian Foundation, who live in Poland. Rzeszów is twinned with Chernihiv in the north of Ukraine, a city that has been three-quarters destroyed and where hundreds of residents have been killed. For several weeks now, the airport at Rzeszów has been guarded by Americans of the 82nd Airborne Division from

Fort Bragg, North Carolina. Aid supplies and large amounts of weaponry for Ukraine are flown in there.

I wander the streets of Rzeszów; it feels like Debrecen or Kaposvár. After passing a beautiful bus station with 1930s lettering and two kebab stalls, I come upon Studio Endorfine and a shop selling statues of the Virgin Mary, right next to a pole dancing studio. On the wall is a poster advertising a 'Burlesque Night', with Ava Afterglow, Nokturna, Swan Savage and Lady Fox. The war is close by here, unlike in Budapest, and so is decadence, a celebration of life at the extremes.

The intention was that the kevlar helmets, the knee pads and the trauma kits would be flown into Poland today or tomorrow from Tampa, Florida, but there's been a delay. A new world of freight transport, consignment notes and customs clearance is revealing itself to me. Thank god I have a friend in New York, Theo, who is helping me with procurement and logistics. I've come to Poland to see whether the Ukrainian aid organization that's going to ensure it all gets to the front looks bona fide.

Outside my hotel I meet Illya, Symon and Lesya. All three are young, in their late twenties, and they look unconventional, likeable. Actually they win my trust without having to say a word. Lesya is small, Illya has jet-black hair and a Cossack moustache, Symon is taller than me and bald, a caring young man. Years ago Symon set up a foundation to help refugees in Poland with language lessons and education. Since 24 February these three have focused most of their attention on Ukrainian refugees, and they've been fundraising to help civilians and the army in Ukraine. They have 150 volunteers assisting them and a fleet of delivery vans that take supplies from Poland to the places in Ukraine where they're needed most. The work and the transport of goods are carried out exclusively by friends, or friends of friends. For deliveries to the army they work in collaboration with the ministry of defence. The trio take me with them to a dimly lit basement for something to eat.

'A new Iron Curtain is coming down and Ukraine wants to be on the western side of that curtain,' says Symon. 'Not on the Mongolian side.'

'The Germans are afraid of offending the Russians,' says Illya. 'It's madness. In the Second World War the Ukrainians suffered more than anyone else, but the Russians have claimed all that suffering for themselves. In the Soviet Union and its satellite states there was no room for Jewish suffering and Jewish victims, and in the same way there were no Ukrainian dead, only dead Soviet heroes. Whereas in fact four million Ukrainians died defeating Nazism.'

'The Germans feel guilty with respect to the Russians,' says Symon. 'They don't realize that in Russia a lot of people are now driving around with bumper stickers saying "All the Way to Berlin!".'

'You should watch Russian television. You won't believe it, but they talk every day about which country's turn it will be after Ukraine,' says Lesya. 'Night after night they go on about the approaching Armageddon. Saying the West needs to be taught a lesson. You should take a look at that idiot Zhirinovsky.' (Zhirinovsky died in April 2022 incidentally, probably of Covid-19.)

'Seventy per cent of Russians think that after Ukraine they'll need to occupy the Baltic States and Poland,' says Symon. 'I don't want to have to fight against the Poles five years from now. Because that's Putin's system; men from the defeated territories have to fight on the front line. Just look at the men from Donbas, at the Chechens, the Belarussians and the Syrians. They all have to fight for Putin.'

'That's why my band is called Endangered Species,' says Illya. 'It's a Ukrainian band I play in, along with my brother. The Russians have derogatory terms for the inhabitants of practically all the countries around them. Here's a list that tells you who they're dehumanizing and making ripe for slaughter.'

Rzeszów, Monday evening, 28 March 2022, 19.00 hours

Derogatory terms

This is the list of insulting Russian terms that Illya noted down for me:

> Chechens and Azerbaijanis: *churka / chukmek* (block, lump; moron), *hadj* (used as a term of abuse for all Arabs)
> Georgians: *gryzun* (a play on words: *Gruzin*, Georgian, sounds like *gryzun*, rodent)
> Ukrainians: *khokhol* (forelock, a reference to the Cossack-style haircut), *malorosy* (Little Russian), *banderovets / banderowiec* (after Stepan Bandera, one of the leaders of the Ukrainian nationalists)
> Belarussians: *bulbash, bulbashi* (potato farmer, from Belarussian *bulba*, potato)
> Balts: *tormoza* (slow; *tormoza* means brakes), *medlennye* (slow)
> Poles: *pshek* (after their strikingly sibilant language)
> Germans: *fritzy, hansy, nazi*
> Dutch: *planokura* (people who smoke a lot of weed)
> Americans: *pindos, pindoses, pindosiya* (capitalists)
> Italians: *makaronniki* (a reference to macaroni)
> French: *lyagushatnik* (frog-eaters)
> Asians: *uzkoglázy* (slitty eyes)
> Jews: *zhyd, zhydovka* (yid)
> Africans: *chernozhopy* (black-arse)

Budapest, Café Zsivágó, Wednesday 6 April 2022

The pain

From 2008 to 2010 I worked on my master's thesis (*From Ballroom to Basement. The Internal Exile of the Hungarian Aristocracy in Transylvania*) at the Central European University in Budapest, before writing a book for the general reader on the same subject (*Comrade Baron*). I interviewed around fifty direct and indirect victims of the socialist dictatorship, and even after the book was published I carried on interviewing them. With varying degrees of openness, former Hungarian and Romanian class enemies told me about what they'd been through in the period 1945–1990. Some said euphemistically, 'It wasn't very nice.'

Several told me in detail what the brutes, trained in Soviet Russia, had done to them. The uncertainty, the endless hours of interrogation, being led down corridors blindfolded, the interminable waiting and listening to the screams of the tortured, the standing against a wall, the being woken time and again, being forced to drink piss, the blows, the electric shocks, being tied to heating pipes, the breaking of fingers and toes, having cigarettes stubbed out on their lips, the insertion and smashing of ampules, being dumped in outer suburbs early in the morning from moving cars.

To gain some kind of understanding of my Hungarian friends and family, of Hungary and Eastern Europe – its psyche and its paranoia – I immersed myself in the methods of Soviet terror. I did so by means of those interviews and by reading literature. In 2012 I developed a heart condition and in 2013 I needed surgery. At the risk of being thought a hysteric: it sometimes struck me that there might be a connection between my heart problems and the miserable stories I'd been absorbing like a sponge for four years. What must it be like for the people who actually went through all that? For their children and grandchildren?

Keeping people in uncertainty, sowing division, denial, lies, manipulations, intimidations great and small, death lists, disappearances, mass deportations, torture, rape, murder, a complete disrespect for human life: there is so much in the operational procedures of Putin and his cronies in Ukraine in 2022 that I recognize from the methods set down by Dzerzhinsky in 1917 and applied in the post-war years in all the Soviet satellite states. That's what makes it so important for me to write about all this.

Felix Dzerzhinsky, the founder of the Cheka, the apparatus of terror under Lenin, had himself been imprisoned and tortured earlier in his life by the tsarist secret police. He was familiar with the lashes of the whip. Pain and suffering are passed down from generation to generation.

What was taught to the torturers in the Soviet Union's satellite states at the Dzerzhinsky Academy shortly after the Second World War is now part of the toolkit of Putin and the people around him. It's being unleashed on the Ukrainians. It's a terrible continuum. I don't want to live in a world ruled by terror. And that reality is closer than many of us, especially in the West, realize. In fact large numbers of people in Western Europe believe Putin and his lies, partly or wholly, and parrot what he says. Even people in my immediate surroundings.

Russian propaganda is effective. It makes people have doubts, about everything. Which is precisely the intention. It paints a picture in which the West, the United States in particular, has provoked this war and Russia had no choice but to react in the way it has. The comparison made is that the US would object if China placed missiles on the US border or in some way or other took over Mexico, occupied it or armed it. Well, for a start, China didn't seize the industrial heart, the most important harbour and ten to fifteen per cent of Mexican territory by force eight years ago. The comparison fails there already. Russia has invaded an independent, democratic country and is trying to subdue it with the most brutal violence. Ukraine does not want to be ruled by Putin's rod of iron and its accompanying nepotism. It wants to be free. That's what everyone, young and old, tells me. Russia is meanwhile trying to give the impression that it's reasonable, that it's a country like all other countries in Europe, while in fact it has reverted to the years of Stalinism, so that at any moment you can be hauled off to god knows where.

Part II
And beyond...

Budapest, Café Zsivágó, Wednesday 6 April 2022

God called me

I'm in Café Zsivágó in Budapest, sitting with the back of my head resting against a samovar and reading WhatsApp messages from Neal. He's joined the Ukrainian Foreign Legion, a force set up in 2014 by a former officer of the Georgian army, Mamuka Mamulashvili, to enable foreigners to fight against separatists and Russian forces in Donetsk and Luhansk. In the eight years since then, thousands of foreign fighters have joined. So Ukrainian army recruiters have gained plenty of experience in distinguishing between proper soldiers and foolhardy adventurers or the mentally unstable. There are now an estimated 6,000 foreign fighters in Ukraine, including Georgians, Belarusians, Swedes and Americans.

For our meeting in Medyka, Neal wore black thermal clothing and a bullet-proof vest out of which stuck a pencil with an eraser on the end. He had a two-day growth of beard, slight shadows under his eyes, and gave the impression of being a serious man, not a romantic in search of an ill-considered adventure in Ukraine. He's married with daughters aged twenty-four and twenty-two and a twelve-year-old son. He owns a landscaping company in Alabama, which he has left temporarily in the hands of a friend. He decided to leave for Ukraine when he saw the images of the hospital in Mariupol that was bombed, and he planned to be gone for six months. His wife was less than enthusiastic. Eventually, after some marital negotiation, she consented. 'If you go for three months or less and then come back, I'll support you unconditionally. Three months mind, and not a day longer.'

It was a week before I heard anything from him. I could imagine him never getting in touch again. Medyka is one of the border crossings where a lot of American military materiel comes into Ukraine, and it's more than possible that people hang around there noting down everything they see passing through before reporting back to the Kremlin. It seemed logical that Neal would be distrustful of me, but from the start he was pleasant and expansive, especially after we talked about hunting. He's a deer hunter, like my brothers, father, grandfather, uncles and cousins.

I sent him a photo of myself with my sons in the snow, hunting pheasant in the woods of Somogy, really the only type of hunt I occasionally take part in, once every year or two – an armed walk in the freezing cold lasting six or seven hours, shooting, plucking and eating two or three pheasants. In the photo we look like a band of poorly armed partisans, with shotguns and fur hats, the kind of daft adventurers they're trying to screen out of the Ukrainian Foreign Legion.

We agreed that he would keep me appraised of his experiences in Ukraine, without sharing any locations or details that could cause danger to him or to others. He's recently started sending me messages almost daily. At first things looked bad, especially as regards weapons for the legionnaires.

In the training camps they're good at weeding out inexperienced or poorly trained fighters, who are thanked and driven back to the border. They amount to around thirty per cent of all those who sign up. The biggest problem are people who are in trouble at home, are suffering the pain of romantic break-ups, have nothing to go back to, or are struggling with depression and using this conflict to commit glorious suicide. One of the more serious legionnaires Neal met refused for that reason to go to the front. He didn't want to be surrounded by amateurs who were tired of life.

'There are American and British troops, Canadians, even a few Japanese,' Neal writes to me. 'Most arrive here in groups and don't let outsiders into their sub-unit, English speaking or not. Even the well-trained groups get limited weaponry and so, to their exasperation, are largely ineffective. An American in Kyiv told me that foreigners don't get the right equipment and are sent out on impossible missions.'

They are given no communications equipment, so there's no way to call for support if they get into difficulties. Neal tells me that the Americans are still refusing to fight in the front line as a result. They don't even get the full complement of ammunition for a fighting unit.

'I heard from another legionnaire that they were sent out on patrol without anti-tank weapons in an area crawling with tanks. I still have to investigate how much of that is true, but reports like that don't help if they want foreigners to come and fight. Most arrive here with grand ide-

as about unlimited supplies of guns and ammunition, and the expectation that they can set to work right away. When they're fobbed off with AK-47s that don't shoot straight, they leave.'

It has got the legionnaires a reputation for being unreliable, so the Ukrainians deny them their best weapons. Because of this vicious circle of mutual disappointment, the Legion has not been very effective so far.

'The Ukrainian troops are good at defence, but they lack proper knowledge of how to mount a successful attack. We, the foreigners, all know how to put tanks out of action. That's the core of our military doctrine and we've been trained in it since the Cold War. In the Foreign Legion the Ukrainians have the best-trained antitank soldiers in the world available to them, but they palm them off with rusty Kalashnikovs and thirty rounds of ammunition. It's a shame, for everyone. A trained fighter without the right weapons might just as well go and peel potatoes. We've come here to fight, not to serve as target practice for the enemy.'

These messages are from the first few days. Neal is in Lviv or Kyiv. I never ask his location. I prefer to hear things after a delay, as I have my whole life – as befits a writer. He hasn't been on a mission yet. These are stories he's hearing from frustrated foreigners.

'The point when I decided to come was on the tenth day of the war. A hospital in Mariupol where women were giving birth was deliberately subjected to an attack from the air. That was the breaking point for me. An attack on the sacred pregnant woman, whom we've learned to venerate and respect as the giver of life. To me that wasn't just a crime against humanity, it was a crime against God. God's voice shouted out that I must do something. So I came here. I couldn't watch helplessly any longer; that would be wrong. God called me.'

Budapest, Friday afternoon, 15 April 2022, 17.00 hours

Empty-nest syndrome in a time of war

One evening a few weeks ago, when Alexey, Yeva and their children were still staying in our guesthouse, Alexey apped that he needed to tell me something. He hurried over through the night in a thick coat. He'd been given the coat by Julia, the old woman from the village who gardens for us. Julia lives in a small house made of loam, she has a weather-beaten face like an elderly Native American, she's poor as a church mouse, and she has a sly sense of humour. Unfortunately her Hungarian is so slangy that I understand less than half of what she's saying. One morning she beckoned to me from the garden. On the wall around the vegetable plot sat a bag made of thin plastic. Julia lifted it into the air, her fingernails ingrained with black dirt. She nodded towards the guesthouse and said it was for the Ukrainians. I took the bundle of old clothes to Yeva. From then on, Alexey sat by the fire on the veranda day and night wearing a warm coat that fitted perfectly, rolling cigarettes, drinking cold red wine, and phoning and apping in search of a future in Europe.

The very first time that Yeva and Alexey invited Ilona and me to come and eat with them in the guesthouse, Alexey showed us the palms of his hands and said, 'I've got hands, I've got legs, I want to work!' He immediately started fixing all sorts of things in the guesthouse and putting up coat hooks while the three boys – Alexey usually referred to them lovingly as 'my tigers' – explored the surroundings and rolled down the hills. Alexey and Yeva definitely didn't want to go to Germany; the Germans were too strict, too formal, too disciplined. (The same applied to the other two Ukrainian families that stayed with us, incidentally: they didn't want to go to Germany either, perhaps partly because Germany was detested for its cowardly, indolent stance at the start of this war). Yeva and Alexey dreamed of France or Italy.

Alexey came into our house with his coat flapping, excited. He wanted to discuss something with me. We went to sit at the kitchen table. There had been two options and he'd made his choice. One was an Italian company he'd worked for before, when he built an air traffic control system for Milan Malpensa Airport. Eager to attract this experienced Ukrainian engineer, they were offering a total package: residence permits and accommodation would be arranged, along with schools for the children and a good contract for three years. Alexey had told me about that opportunity a few nights before, under the wide starry sky by the fire. It seemed to me a golden opportunity. It sounded solid, and Milan struck me as far from a bad place to live.

The other option was being offered by a Ukrainian friend, married to an Italian man, who owned one of the best steak restaurants in France, called Lucky You, Beef and Seafood. Alexey could go there to work in the kitchen. Perhaps he might start out at the Lucky You Saloon, the prestigious restaurant's rather simpler little brother, located nearby. As a child I went to Cannes every summer and mainly remember the marina with the oligarchs' boats and the Rolls-Royce Silver Shadows on the quayside. I found it insufferably swanky (I prefer to wander in the woods) and vulgar. It seemed to me an uncertain future for Alexey and his family, starting again as a kitchen assistant.

But he'd passed up the security of doing what he'd been doing for the past ten years and was good at. He'd worked for his father's company building air traffic control installations in Italy, Ukraine, Belarus, Lithuania, Uzbekistan and Kazakhstan, among other places. He'd been in charge, and he'd thoroughly mastered both the theory and the practice. I'd seen how good he was with his hands as he refurbished things in and around the guesthouse.

'Jaap, if I sign the contract with the Italians, they'll send me to Africa, to Libya, Liberia, Ethiopia, to war zones to rebuild what's been destroyed. I won't be able to refuse and I'll be away for months. Yeva doesn't want that any longer, and neither do I. For years I've worked day and night in difficult countries and not seen my children for months on end. I want to be with them. With our friends in Cannes I can learn something new, give my life a new direction. I like roasting and grilling meat. I like cooking. It's a huge opportunity, a chance I'd never have had without the war. I can change my life.'

Alexey sat with his likeable, unshaven face close to mine. He talked wildly with his hands, and his torrent of words, passionate and not always logical, seemed to me a good fit for the Mediterranean. I realized that his choice was the right one.

'We're leaving tomorrow morning. Our friend has found an apartment for us through the city council,' Alexey said. 'If we don't get there within a few days, they'll give it to someone else. On the way there we'll have to get documents for Yeva's mother in Italy. She doesn't have a passport. She only came to Kyiv for a few days to look after the boys and she brought nothing with her. We're leaving at nine in the morning, so that we'll be in time to arrange the paperwork.'

I didn't want to lose my Ukrainians. We'd sat together by the fire almost every evening. After a day of writing in my study I appreciated the company. It was perfect for me; I could visit Alexey and Yeva whenever I felt like it. During the day the blissful sound of children's voices rose up from the valley. Misha, Gosha and Jenya foraged about, played wildly with our dogs for hours, tirelessly, and had games of floor hockey with the children from the other Ukrainian family, cheering whenever somebody scored. That conviviality in the background gave me as much pleasure as the chirping of my favourite birds in May. The next day the sky was a brilliant blue. Both families left. A final quick cup of coffee on our patio, hugs all round and they were gone.

I walked across our land, like a war veteran visiting a silenced battlefield. It was horribly quiet. Everywhere I could see traces of the three little boys: sticks and branches that had been dragged around, probably used as weapons in a battle, lay abandoned at random places; the door to one of the potato cellars was wide open, so they must have gone down the narrow brick steps into the dark and the cobwebs. Chunks of cement had been broken off the steep buttresses of the cowshed, so they must have climbed up those. At the little tennis pavilion, the terrace was barricaded with chairs and upturned tables. It brought a grin to my face and it made me melancholy. It was as if my own three sons had just left home. Dammit, I had empty-nest syndrome, caused by a Ukrainian family off to embark on its new future.

Amsterdam, Sunday evening, 17 April 2022, 23.00 hours

Modern Spartans – the stories of survivors in Bucha

'Understand me, Jaap: these people bring me to my knees. I'm a hardened war veteran. I've fought in Iraq, I've been in Afghanistan, I've seen everything. But what I've seen and heard in this city I've never experienced before. The level of cruelty and suffering bears no comparison to anything I've known in other wars.'

I'd just arrived in Amsterdam. It was a beautiful day, a golden spring day, the sun glancing off the water, the canals full of boats, youngsters sitting on the canal walls, the streets thronged by cyclists, people on the café terraces turning their faces to the sun. At the same moment, it was raining in Bucha and police, civilians and relatives were searching the mud for the missing. Covered pits were reopened and bodies pulled out of the groundwater. Everything was being precisely recorded. Police officers took photos and fingerprints, trying to identify the dead and where possible to reconstruct the massacres. They were then reburied, sometimes in the presence of a small clutch of relatives. So far, more than five hundred bodies have been found, about eighty per cent of them obviously killed by an act of violence, often a bullet to the back of the head. The Russian troops went door to door in apartment blocks, shot the locks open, dragged the women to the cellars, looted the flats and usually killed the men.

Neal is in Bucha. He writes to me that he's helping to collect evidence of the war crimes that have been commit-

ted there. He takes photographs, interviews people and supports them. He sends me the stories and impressions that he can't very well share with his wife and children. I'm shown videos of a destroyed Bucha and captioned photos of him with survivors. He looks shaken. He's wearing a camouflage boonie hat. He sends me his stories late in the evening, two of which I note here. The first is a tale of forgiveness. The second is a cruel report. Anyone who thinks they might not be able to cope with that would do well to skip the rest of this chapter.

The first photo is taken through a doorway. A fat elderly woman sits on the edge of a bed, wrapped in rags and exhausted, supporting herself with her hands braced on her knees. Behind her is a window. There's too much backlight to see her properly. Neal gives his report in short sentences.

'I don't know what her name is. Everybody calls her Babushka, meaning "grandmother". She's ninety-seven. She was living alone. Her husband died fifteen years ago. He was a veteran of the Soviet army. He served for twenty-five years. She lived in the house that they built together fifty-five years ago. It was destroyed by a Russian tank. Babushka was pulled out of the rubble. All her possessions were destroyed, everything they'd built up during their lives, but she was still alive. She doesn't even have any photos of her husband left, nor of her only son. Her son served in the Soviet army too. He died in Afghanistan. He was a helicopter pilot and he was killed by a Stinger, supplied to the Mujahidin by the Americans. She knows that, yet she kisses me on both cheeks as a greeting and says I don't need to worry, that she doesn't blame me. Babushka decided that God still loved her when she realized she was still alive. She doesn't want her heart to be clouded by hatred on her dying day. She forgives the Russians for what they have done. She can forgive them because she knows that one day they will have to answer to God.'

In the second photo Neal is standing on a stretch of thin grass, with leafless trees and wild pines in the background. It looks cold. He's holding a fragment of a rocket in both hands. Next to him stands a young woman with her hands in the pockets of a thick grass-green parka, the fur-lined hood pulled over her head. Behind them to the left you can see a bit of the wall of a light-green house, a kind of school building. To the right, a steel beam lies where it has fallen. In a third photo the woman is standing with a little girl in front of the doorway of the same building, to judge by its light-green door. The woman and the girl are holding each other's hands. The girl has blonde hair and looks about six years old.

'This is Yulia. The metal I'm holding is a fragment of the Russian rocket or bomb that destroyed her house. Her husband was an ordinary soldier in the Ukrainian army. He fought near Mariupol, as part of an attempt to liberate the city from the outside. He had a week's leave and came home to spend time with his wife and six-year-old daughter, five days before the Russians took Bucha. After their house was destroyed, they tried to flee. Their car was stopped by Russian soldiers and they were detained. His uniform was in the car. All three were taken and tied up. The husband was tortured in front of Yulia and her daughter. They cut off his fingers and stuffed them into the girl's mouth. Then they tried to rape the six-year-old girl. Yulia said that they first tried to rape her daughter from the front. The girl screamed. They couldn't do it so they raped her from behind, three men.

'There were three of them in that room. Yulia's husband, mutilated by torture, was first forced to see his daughter being raped and then had to watch helplessly as Yulia herself was raped, for days on end, by one Russian soldier after another. Soldiers came, raped her and beat her husband, the father of their daughter, until he was unconscious, time and again. In the end, when it was time for the Russians to move on, they came into the room, shot her husband in the head and left.'

I look at the photo again. The girl looks sweet and tough, the kind of little girl who climbs trees. In front of Yulia and her daughter are two big sheepdogs. The girl is fearlessly stroking the head of one of them, holding her mother firmly with the other hand.

'It was a week before Yulia could walk. The little girl needed stitches. Her mother says it's a blessing she's so young. The doctors told her the girl will be okay. If she'd been a bit older she'd have been made infertile for life. I didn't ask Yulia any questions, I just let her talk. She speaks good English and she managed to explain to me what had been done to her family. Despite all this, Jaap, she was able to embrace me. She was able to put her arms around a man, rest her head on my shoulder and thank me for coming.

'What happened here is happening all over Ukraine. I don't know if we can stop anything here. These people have superhuman strength, a strength I've never come upon in my life before. Their resilience is immeasurable. God really does live in these people. I'm no hero, Jaap. These people are heroes. And the incredible thing is that Yulia and her daughter can still smile. I don't know how, but they can. These people are unbelievable. I'll respect them till the day I go to my grave. The halls of Walhalla are filled with Ukrainians. They are the modern Spartans.'

Naples, Gran Caffè Gambrinus, Tuesday 10 May 2022

Karma is a cruel thing

Within myself I notice a puzzling side-effect of the war, a side of me that I'm almost ashamed of but can't deny: it's given me a pleasant sense of urgency and direction. I may have mentioned that my brothers and I, five of us brought up by a lone mother in sometimes stressful circumstances, flourish in times of tension and conflict. We're not averse to switching to survival mode. I can only speak for myself, but it's as if it makes me fall into place. Along with my wife's family, we're still taking in Ukrainian families in Hungary, and with Dutch and Ukrainian friends I'm delivering aid to the front line.

It's evening and I'm sitting in the beautiful Gran Caffè Gambrinus in Naples, with its marble floor and gold on the ceilings, where the waiters wear black aprons and dinner jackets and where Oscar Wilde, Gabriele D'Annunzio and Ernest Hemingway once drank coffee, or something stronger. The doors are wide open and a sea breeze blows in. Mopeds and scooters tear past outside and church bells are ringing.

Over the past two weeks it's become clear that the war in Ukraine is overflowing its banks further than ever. It's been doing that for a while, noticeably with the torrents of refugees and the pleasing confiscations of properties belonging to megalomaniac Russian tycoons and Putin's henchmen, but now the war is fanning out further, with more areas and countries being affected. In Transnistria two telephone masts were brought down by explosions, in France glass-fibre cables were cut and Russia uncoupled Bulgaria and Poland from its gas supplies. I heard through a high-ranking officer in the British army that there are fears the Russians will cut the fibre-optic internet cables on the bottom of the ocean between the United States and Europe.

In the same period the United States announced that its intelligence enabled the Ukrainians to sink the Russian flagship the *Moskva*. Hundreds of kilometres beyond the Ukrainian border with Russia, important oil and ammunition storage places blew up, in Briansk, Kursk, Voronezh and Belgorod. To the north of Moscow, in Korolyov and Tver, institutions that are strategically important for the development of rockets mysteriously caught fire – the Russian state news agency ITAR-TASS reported that the fire was caused by electrical faults. A few hours later in Kineshma, more than nine hundred kilometres beyond the Ukrainian border, the largest chemicals factory for the production of Russia's rocket fuel burned to the ground. Ukraine neither confirms nor denies involvement in the inexplicable fires deep inside Russia. Myhailo Podolyak, an adviser to President Zelensky, wrote on Telegram that it was probably a matter of karma: 'Karma is a cruel thing.'

While the conflict extends its reach, Avril Haines, America's director of national intelligence, warns that Putin is making preparations for a long war, hoping for declining interest and aid on the part of the United States and Europe, and for the economic collapse of Ukraine. I see the war everywhere. Even in Naples it follows me. In front of my nose, in the haze of the grey-blue bay, is a vast aircraft carrier, the USS *Harry S. Truman*. It has just sailed in, surrounded by a fleet of navy ships and submarines. It has at least fifty planes on its deck, in battle order, ready to take off. This too is a means of communication.

I exchange messages with Neal daily. He's now in the east, training Ukrainians to be snipers. I'm also in touch, again almost daily, with two young Ukrainians, Illya in Poland and Dima in Kyiv, who are making sure that the things I manage to get hold of end up at the right places on the front line. Chance has made a fundraiser of me.

At the border I saw many large aid organizations caring for refugees. Britain and the US have sent weapons. There's a yawning gap between those two forms of aid. The airfields in Poland are now being inundated with weapons and other things for Ukraine, including the supplies I've been able to acquire thanks to the financial help of friends, to protect both civilians and soldiers. The ministry of defence in Kyiv decides what goes where. Illya and Dima coordinate the transport to the front, using delivery vans. They let me know how many helmets are going to Chernihiv, Korosten, Donetsk, Kharkiv or Mariupol.

As soon as a van arrives at its destination, Illya sends me a photo by Telegram of the supplies in the hands of the troops, often with the men's faces made unrecognizable. Illya was unable to tell me how, but supplies were delivered to the marines in Mariupol until the last possible moment (recently he told me that at that point, of every three vans that left for Mariupol, an average of only one came back).

It's a small contribution, but satisfying. On the day when Vladimir Putin announced with a straight face on Russian state television that the steel factory in Mariupol would not be stormed and he was having it sealed, so that 'not even a fly can get through', a Dutch consignment of trauma packs, kevlar helmets and night vision goggles was smuggled in to the marines in Azovstal.

Somogy, Saturday afternoon, 14 May 2022, 16.00 hours

On the killing of generals

'I recall the pain and fear on the faces of those who came out from cover. I hear their screams for help. I hear the terror in the throats of the others who had sought cover. I hear the tremble in their voices. I see the fear in their eyes. I observe the shock when my bullet goes through them. I remember everything, every detail of everyone I've killed. It's like having sex with a woman you love. It never leaves you.'

The Ukrainian army claims that since the start of the war, it has killed eight Russian generals and fifty colonels. That number is extraordinarily high and it's causing problems for the Russian army. The deaths of several commanding officers have been confirmed by the Russians. They include Major General Vitaly Gerasimov, the nephew of one of Russia's highest ranking military men, as well as Major General Vladimir Petrovich Frolov, Major General Oleg Mityaev, Major General Andrei Kolesnikov and Captain Andrei Pali, deputy commander of the Black Sea Fleet.

The Chechen general Magomed Tushayev, leader of the 141st motorized regiment of the Chechen National Guard, led the assassination squad sent to kill President Zelensky. Following the American example, the Chechens in the brigade had been given packs of cards with photos of the members of the Ukrainian leadership who needed to be taken out. Magomed Tushayev was one of President Ramzan Kadyrov's faithful comrades and had been responsible for the torture and extermination of Chechnya's LGBTQ+ community. On the third day of the war, Tushayev was caught in a trap in Hostomel, close to Kyiv, along with his men, by a Ukrainian elite unit. Some Russian and Chechen sources deny that Magomed Tushayev was killed.

The Russian army is extremely hierarchical in its organization, with little room for initiative or decision-making by the lower ranks. The Ukrainian army, by contrast, has been trained in recent years to operate in small independent groups after the Western model. According to *The Wall Street Journal*, Ukraine has set up a special unit to track down generals and other senior officers and kill them. The Russian officers often use unencrypted radio and telephone communications, so they can be located easily.

A Ukrainian musician friend of mine told me how in the early days of the war he was in a private WhatsApp group where several sound technicians managed to crack Russian military radio traffic. With the members of his band he set up a duty roster so that they could listen to and record conversations 24/7. He passed everything on to Ukrainian military intelligence. The Russians also communicated on Russian radio in Morse code. In the future he wants to compose a number based on that sound, the sound of the Russian Morse messages he's recorded.

Most Russian tanks, which give themselves away at night because the crews leave the engines running to stay warm, are blown up using Bayraktar drones and Javelin missiles. Because of a lack of motivation among the ordinary troops, high-ranking officers in the Russian army often have to go up to the forward lines to issue instructions and orders. There they are vulnerable. Generals and colonels are shelled and sometimes killed by snipers, as happened to Major General Andrei Sukhovestky.

Neal told me how he went into Russian-occupied territory repeatedly for forty-eight hours at a time to seek out and kill Russians. I asked him how that worked. We'd earli-

er corresponded about the deer hunt. I'm familiar with the stalking of deer on the open hillsides of Scotland. You see them from a great distance and try to sneak closer, ideally as close as 150 metres, to get a proper shot at them. If you're a good marksman, 200 metres is close enough, but the bullet drops quickly. You adjust your sights to the distance, to correct for the fall of the bullet. The approach is not easy, since deer have sharp noses and excellent hearing. The older hinds in particular are watchful.

'My son and I hunt deer every year in Alabama,' Neal writes. 'In our community we hunt for food, not for trophies. Deer aren't allowed in the area; on a terrain of fifteen square kilometres, almost three thousand of them need to be shot. Every year my son and I shoot between five and eight deer. Five go into the freezer and I give the rest to friends and family. For me and my son it's all about the time we spend together in the outdoors, respecting what nature can give us, as long as we have sufficient patience. There's a lot to learn in the woods. About our world, about nature and animals, and above all about yourself.'

I was curious whether the Ukrainian Foreign Legion was helping to liquidate colonels and generals, so I asked Neal.

'I once got a specific commission. All the other times were chances that offered themselves: junior officers, lieutenants. We work as a small team. We always get a lift to a place where Russians are known to be. We go into Russian-held territory and after forty-eight hours we're picked up at an agreed place. We walk for miles and usually make the hit shortly before it gets dark. I don't have night vision goggles, but the rifle does have a residual light amplifier. I shoot a .308 with a silencer at a distance of five or six hundred metres. I bought the rifle with my own money, here in Ukraine.

'I travel mainly with Ukrainians and I live like them, dress like them. I'm not driven by vengeance, but I can imagine that does play a role with the Ukrainians. The Russians are doing terrible things here. We try to kill them while they're walking around. They always run away and never try to find us. I take aim at the leaders. The sergeants, the officers. They're easy to pick out; an officer is generally one old man surrounded by a group of youths. I shoot him first. I try to spare the lads. I don't kill them unless I have to. They're children, between eighteen and twenty years old.

'When they're on the road I shoot at the drivers, who drive with their heads sticking out of the tank. A tank makes such a racket that they've no idea where the shot came from. Three days ago I shot four of them.

'We're always alone. The risky thing is that we don't have any support and can't call for any. We have a limit-

ed amount of ammunition with us, so we're not allowed to get into a firefight. After firing the shot we pull a thermal blanket over us and keep still, in case they send helicopters with special optics out after us. The blanket blocks our infrared signature. After a while we move, usually a kilometre or two.

'We go deep into Russian-held territory and strike terror into them. We move mainly at night, stay a good distance away and keep quiet. We hide in daylight hours. The night is our friend. As is the forest. It's like hunting, only for a different prey. My hunting experience helps. Imagine a deaf, blind deer with no sense of smell. That's what it's like hunting Russians. As soon as we get ourselves into a firing position, we move very slowly. Always from cover and never in the open field. Sometimes I shoot from a tall building. The silencer helps. It keeps us quiet and hides the flame of the shot.

'Once I was sent out to kill a senior officer. He'd been located and my instructions were to take out the target if he put in an appearance. If he didn't turn up within two days, I could shoot whatever I liked. I was lucky; the target soon showed himself. We had a photo of him with us. I thought it was him; my spotter was a hundred per cent certain. The officer was at a Russian advance post. He was there to determine the strategy on the front line. I don't know whether he was a general or a colonel. He wasn't wearing a uniform at the moment of his death.'

In the train from Dombóvár to Budapest, Monday 16 May 2022

The Cossack mentality

'Good evening. My call sign is Phaeton. I work at the Technology Museum in Zaporizhzhia. I'm a gunsmith. Since 2014 the museum has been helping the Ukrainian army to refurbish old weapons. On 24 February I reported to the army as a volunteer. Soon I'll have to go to the front. Neal is my combat instructor. I'm a repair, maintenance and operation instructor. I can refurbish and set up a Maxim gun, for example. They were produced in great numbers in 1942–45, but before then too. Russia was already using the Maxim gun in 1905, in the Russo-Japanese War.'

Neal has introduced me to a gunsmith who restores hundred-year-old machine guns from museums and army warehouses, which then go straight to the front. He's a rather ageing, corpulent man. I know that from a photo he sent me via Telegram with his little daughter and a huge fish they'd caught. The fish is as big as his daughter. Neal has asked whether I can arrange for a bullet-proof vest for his friend, level four, which protects against 30-06 ammunition, and get it to Zaporizhzhia. That's where the man is. A place that has mythical significance for many Ukrainians.

Zaporizhzhia – the name means 'beyond the rapids' – was traditionally the centre of the Ukrainian Cossacks, Illya tells me. The Cossack fort was at the rapids in the Dnipro River. Illya, a band leader, graphic designer, violinist and guitarist, who helps me from Poland with the organization of transport for supplies to the front, has become a friend.

I've visited him in Poland and he's been to see me in the Netherlands. He looks like a Cossack, with a big handlebar moustache and black hair closely shaven at the sides. He's an extraordinary young man, musically gifted, and I can talk to him about Le Corbusier, Béla Bartók and Jack Kerouac. He's highly educated, certainly for his age. He's twenty-eight. Some time ago we agreed by phone which night vision goggles it would be best to buy for the marines in Mariupol. The Azovstal factory had not yet been lost. 'How splendid,' I said, 'that a violinist and a writer are discussing which night vision goggles are best for the front.'

Illya hasn't shaved his head completely leaving a single long lock of black hair. That would be perfectly in keeping with the Cossack style, known as the *oseledets*. The Cossack look has been in vogue ever since the start of the war in 2014. Illya tells me that the Cossack mentality, the quest for individual freedom, forms the core of the resistance to the Russians. He says it represents the fundamental distinction between the Ukrainians and the Russians, who have remained a people of serfs and have never truly freed themselves from the bottom up. One tsar followed another. The Cossack communities originated from escaped serfs, men determined to be free. The Cossack myth of the ultimate free-spirited man is an important ingredient of the current Ukrainian self-image.

The gunsmith in Zaporizhzhia also refurbishes the machine guns that are unscrewed from Russian tanks and armoured cars. That fits with the Cossack tradition of inventive, nomadic raiders. As well as pillaging, fishing, poaching and small-scale farming, they engaged in the freeing of slave transports. For centuries the Crimean Tatars provided manpower to the Ottomans. Slave trading was the engine that powered the growth of the Ottoman Empire. The Crimean Tatars stole slaves in Ukraine and Russia, especially in years when the harvest was bad. They're estimated to have hauled some two to three million slaves out of these parts. The sons and daughters of the peasants of Galicia and Volhynia were sold at the slave markets of Caffa (present-day Feodosia in the Crimea) and Constantinople. The men ended up in the galleys or as Janissaries, the women as kitchen slaves or as concubines in Konya or Cairo.

The endless flatlands of the Carpathian basin slide slowly past me. I read the messages from the gunsmith in

the train from Budapest to Dombóvár. I've made all kinds of friends because of this war, people I would never have met but for the Russian invasion. Like practically everyone, I live in my own bubble. Because I travel between two countries, the bubbles are reasonably varied: Hungarian villagers, gamekeepers, foresters and tractor drivers, distinguished Eastern Europeans, a varied mix of people living in Budapest: ambassadors, historians, spies, artists, network specialists and idlers. Then there's a very extensive and lively Dutch and Hungarian family, neighbours in Twente, old friends from my student days, a few Dutch writers, publishers, filmmakers and actresses, plus architects, carpenters, plumbers and painters in Hungary and the Netherlands, whom I've worked with for years and who have become friends. Now a few Ukrainians have been added to them.

That the Ukrainians are getting weapons from museums and old transit warehouses, and removing them from blown-up tanks, shows how short of supplies they are. But they're not the only ones. I understand from messages from the gunsmith in Zaporizhzhia that the Russians are using old weaponry too.

'Near Kyiv our lads got hold of a machine gun from 1914, in use by the Russians.'

The train glides across the landscape. The fields have been ploughed and levelled for the summer crops. Soon maize, sunflowers and lucerne will be sown. There are black raked seedbeds as far as the eye can see, with light patches, like hair where the dye is growing out. Hawthorn trees blossom soft white.

'My favourite is and remains the Maxim gun. If there's enough water available, it can fire 36,000 times without stopping, no problem. There's a reservoir around the barrel with an opening at the top. If you just keep topping it up with water, you can carry on shooting without overheating. I've restored about twenty Maxims now. They're in use all along the front line and there's enough ammunition for them, the same belts as for the PK machine gun. A B32 armour-piercing round fired from a Maxim penetrates the armour of the Soviet APCs. It's an extremely accurate weapon. If you hit your target at a kilometre and a half, or even two kilometres, you can just pull the trigger and all the bullets will fly to the same point of impact. There's only one disadvantage: the thing weighs seventy kilos.'

A skinny hare zigzags away from the train, ears flat. Past ditches and brooks, on wet or sloping land are patches of woodland, still without leaves. Sometimes we pass the edge of a village with low peasant houses, the back gardens stashed full of things that might one day come in useful.

'I'm now working on four machine guns. I don't think I'll get them finished before I go to the front. Three of the four are from burned-out armoured vehicles. It'll be hard to make them work. They're a standard machine gun that's mounted on a lot of Russian materiel, on the T-62, T-64, T-72, T-80 and T-90 for a start. We got two of them from BTR-80 armoured personnel carriers. They're Russian trophies. Very symbolic. Our men will defeat the enemy with Russian weapons.'

Budapest, Saturday 21 May 2022

The Russian method

It's morning and I'm sitting in Budapest in a wobbly seat on the terrace of Café Zsóka, with a cup of coffee, working on this book. At the table next to me, two girls are speaking Russian. One, with grey-blue eyes, looks watchful, the other is slumped in her chair, staring around her, bored. I ask where they're from. Saint Petersburg, they tell me. Rich Russians are rather fond of Budapest, as they are of Vienna. It's neutral, close by, and equipped with all the material attainments of the European Union. The Andrássy út is full of the pricey shops of the Louis Vuitton Moët Hennessy group, which has Russian and Hungarian oligarchs to thank for a substantial part of its local turnover.

The girls have been studying in the EU for six months as part of the Erasmus Programme, one doing languages in the Czech Republic, the other political sciences in Lyon. The political scientist, Valeriya, is on her way home. She comes from a liberal university in Saint Petersburg and proudly shows me the cotton bag of her alma mater. Professors at her university have spoken out against the war, signed petitions and then been sacked.

'It wasn't easy in France, especially at the start of the war. Our bank accounts were frozen, people were hostile when they heard you were Russian, and everywhere I went I had to keep explaining what I thought about the war. In Paris I was verbally abused by a Ukrainian who heard me speaking Russian; he said I should burn in hell for eternity along with my whole family,' says the girl with the grey-blue eyes. 'I'm half Ukrainian, my mother is Ukrainian. I don't support Putin, most Russians don't. Friends of mine have gone on demonstrations. One was arrested and held in a cell for two weeks – no idea whether she'll be allowed to continue her studies. Everything has changed since February. I probably won't be able to study abroad any longer.' That's the Russian method: exclusion and reprisals for all those who refuse to cooperate, and intimidation for everyone else.

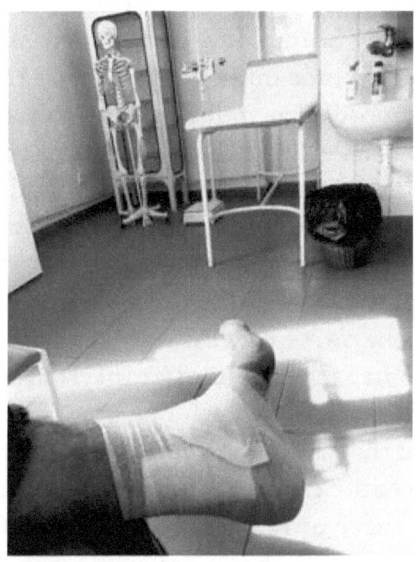

I still come to Budapest for a haircut. I don't dare do that in the countryside. I do go to the local doctor, though, who comes to the village once a week, on a Tuesday morning, and sees his patients beside a wood stove in the village hall. In the space used as a consulting room there's one metal cabinet and a life-sized skeleton, and you can look in through the windows on all sides. A week after a car ran over my foot at the border at Lónya, I visited the travelling doctor. My foot was swollen and purple, and it itched maddeningly. He's an experienced doctor who limits himself to kill or cure remedies, which suits me and usually works well.

But I'm not drawn to kill or cure remedies for my hair. I don't know who the barber is, but many of the villagers are bald, and the young men have closely shaven heads, like footballers. The older women in the village have short perms, often red, the colour of Ecoline, the colour of blood. The herds of sheep in the fields below our house get exactly the same red colour on their backs and haunches in late August from the ram's tupping crayon, indicating which ewes have already been mounted. I suspect the raddle dye and the hair dye come out of the same vat in our village. Which makes more than enough things to avoid, so for fifteen years I've been visiting the same hair salon in Buda. I'm a man of routines, much in the way that Ludwig Wittgenstein would have preferred to eat exactly the same meal every day. That's gone by the wayside; Ilona likes variety.

For the past few years the task has been performed by Gabi, a young hairdresser with jet black hair. Gabi comes from a village in Székelyföld in Transylvania, which gets me on her side straight away. It's an area I know well. The Szeklers are hard-headed and they form a strong community. It's a tribe centuries old, a Hungarian minority in Romania, living in three provinces up against the Carpathians. Gabi's fiancé comes from two villages away. They're soon to be married and I get to hear all about it. She's paying for the wedding herself, since her parents don't have the money. Her father works in the forests. They're inviting three hundred guests. One day they want to move back to Székelyföld together, where Gabi will open her own hairdressing salon.

My Budapest hair salon atones a little for the Treaty of Trianon. It's a reception centre for hairdressers who have their origins in the Hungarian minorities in neighbouring countries. The last time I came, Gabi wasn't here and instead I got a plump girl from Ukraine. Gabi now has three colleagues from Ukraine: Henrietta, Klaudia and Krisztina. None of the three speaks Russian or Ukrainian, only Hungarian. They've lived in Budapest for years. While doing my hair, Gabi tells me that relations between Hungarians, Ukrainians and Russians in Transcarpathia in Ukraine were good until the Maidan revolution in 2013.

'There's a shortage of jobs and food in Transcarpathia. The soil isn't being cultivated,' Gabi says. 'Only the women can work, because the Hungarian men have to hide from the police to avoid being sent to the war. Hungarians from Ukraine can easily get jobs here in the beauty industry. They know the Russian techniques, the Russian method.'

In the evening I go for a drink with Rupert and two English friends, Tara and Philip, who have been living in Hungary for decades. We stand at a tall table and everyone is so disappointed at Hungary's attitude to this war that we prefer not to talk about it. I tell them I'm arranging for night vision goggles, drones, helmets and trauma packs to be sent to the army, and in a few days from now I'm going to be delivering tools. Rupert is just back from Kyiv, having driven a transit van full of night vision equipment and other semi-military supplies there via Poland. Philip says he was walking along Hadrian's Wall in northern England when he spontaneously drove to an army surplus store, bought second-hand bullet-proof vests and knee pads and sent them to Ukraine. I'm pleased that we all seem to have the same ideas and impulses. With the Brits you can win a war. Tara says that last week she accompanied a group of Ukrainian women, wearing folk costumes they'd borrowed from an ethnographic museum in Ukraine, through the streets of Budapest for a pop-up performance on Deák Ferenc Square. The women were sworn and spat at by Hungarians in the street. 'Not once but ten times.'

We shake our heads. Rupert, who has been working from London in corporate investigation for thirty years, with a focus on Eastern Europe and Africa, says that around ten years ago Putin proposed to Poland and Hungary that they should split Ukraine up between them.

'Hungary would get Transcarpathia,' Rupert tells us. 'Poland would get a piece of north-western Ukraine. I don't know the precise date, but the proposal was made to Orbán and Tusk, so it must have been between 2010 and 2014, in preparation for the occupation of the Crimea and Donbas. Putin was trying to spread the guilt as widely as possible in advance. It illustrates how cunning he is. Tusk immediately said no. Nobody knows how Orbán reacted.'

This is part of the Russian method, too: make others your accomplices. It was the essence of the Soviet system, that hall of mirrors founded on corruption and humiliation that nobody could escape. Years ago I studied the treatment of prisoners in Romanian prisons under communism. The programme was based on the ideas of Soviet sociologist and educationalist Anton Marenko and developed by Ludovic Zeller and Boris Grunberg. The subject must be made to understand that he was a *déclassé* and that his only hope of salvation lay in gaining the support of the Party, which he could do only by bringing fellow failures to the path of truth by means of 're-education'. The goal was to destroy the personalities of opponents of the regime. They were given a choice: die or beat up fellow prisoners. New inmates were thrown into the courtyard naked and set upon by other prisoners with iron rods. It was the first step in their 're-education'. As a prisoner you'll never tell anyone what you did to survive.

The way such a system works is interesting. The essence of it is rape and torture. Around that, a sense of menace and circles of complicity spread wider and wider. The guilt you feel can be caused by something small. That you held your tongue when injustice was done right in front of your eyes. That you informed on someone. That you profited from advantages or money gained dishonestly or by corruption. Or that you took the apartment or job of someone who'd disappeared. In Russia under Putin, 're-education' and the whole empire of humiliation that goes with it have been reinstated.

A couple of years ago an inmate of the Saratov Prison copied images of torture and smuggled them out, as evidence of systematic oppression in present-day Russia. Led by the guards, prisoners were ordered to take a fellow prisoner aside, undress him, hold him, beat him and rape him with a broomstick for hours. This was filmed with a bodycam and neatly filed away by order of the directors of the prison, so that the tormented prisoner could be blackmailed later with images of his rape. Courageous witnesses revealed the fact that systematic torture takes place in Russian prisons, such as Saratov and Prison no. 1 in Yaroslavl. Russia has become an empire of humiliation again, as it was in the worst years of Stalinism: humiliate or be humiliated.

Since Putin came to power, the Russian Federation has subdued neighbouring countries (Chechnya) or annexed parts of them, as with Abkhazia and South Ossetia (from Georgia), and Luhansk, Donetsk and the Crimea (from Ukraine). The men of the conquered territories are then given a choice: join the fight against the next enemy or be crushed. Boundless expansion, and the robbery that goes with it, is necessary if corrupt, megalomanic regimes are to hold on to power.

Budapest, Café Csendes, Sunday evening, 22 May 2022

Putin is fantastic!

'Many Russian friends block me on the internet. My mother begs me to stay in Hungary because Orbán is a good friend of Putin's. She keeps saying: Orbán is a great guy, just stay in Hungary. Russian television is extremely positive about the Hungarian premier, that's why. But I want to go to Germany, preferably to Hamburg. My mother is narrow-minded. In Pervomaisk there was a lingerie shop underneath the flat where we lived and you continually had to give it money. She wore the lingerie herself. My father paid for that; he earned a living by renovating houses.'

I'm sitting with Margarita under the coloured lights of Café Csendes in Budapest. She was originally a chef, but she's spent years as a web designer. She now works remotely for a web company in Kyiv, as well as doing voluntary work and giving fitness classes in Budapest along with a friend from Irpin. The friend from Irpin fled a day after Margarita and hid for two weeks in the mountains of Transcarpathia before coming to Budapest. Margarita is tall and slim. She's thirty-nine. She fled Kyiv on 2 March, six days after the start of the war. She seems to feel a bit guilty for being safe in Budapest. 'I would like to fight. My soul is Ukrainian. Now I do voluntary work, send money to the army and support our military forces that way. My mother says, "Putin is fantastic!"' Margarita gives me a tormented look. 'I don't want to hear that bullshit any longer. She sends messages from Moscow about American biological laboratories in Ukraine. My brother lives in Moscow; he married there years ago. When the trouble in Donbas began, in 2014, my mother went to join him there. My brother is three years older than me. After his time in the army he started renovating houses, just like my father, and he's good at it. I always dreamed, even as a little girl, of having a big brother who would protect me. A brother I could be proud of. But it's more the other way around. I've always had to protect him. He goes along with the herd, preferring not to think too much. My brother and mother believe the Russian news. I don't. I prefer the truth.'

I've only just met Margarita. She looks tense. I and some friends of mine in Budapest, mostly a lot of Brits and gradually more and more Ukrainians, set up a WhatsApp group in early March to help refugees. Will and Peter of Brody House took the initiative and turned one of their rooms in Budapest into a Ukrainian meeting place. In the App group, questions are asked every day: about homes for families that urgently need accommodation, schools for children, childcare for toddlers, work for mothers, or problems with paperwork. I've written that I'll be going to Kyiv in three days from now, asking whether I could transport things there or back for people. Margarita responded in the App. I don't yet know what she wants.

'From 2014 onwards my mother visited me regularly, sometimes staying for weeks. Even then she didn't want to watch any Ukrainian television programmes. She held the Ukrainian government responsible for the war in Donbas and for ruining her life. When the subject of politics came up she'd have rages, losing her mind. She cursed everything Ukrainian, the history, the culture, the language, the people. I was born in Pervomaisk in Donbas and grew up there. I'm half Ukrainian, half Russian, partly Jewish.'

She plucks at her skirt. The waiter brings our drinks.

'My father, by contrast, felt a hundred per cent Ukrainian. Even though his father was from Russia, from Voronezh. His mother, my grandmother, was from Rivne, a town in north-western Ukraine. He used to say, "You were born in Ukraine, you'll die in Ukraine." I thought about staying in Ukraine and signing up for the army. I'm a good shot; my father taught me that. He used to take me with him to the rifle range. He died three years ago.'

I ask her why she'd decided not to join the army.

'As a girl I dreamed of becoming a doctor, but I can't stand the sight of blood, so I gave up that idea. That's why the army might not be ideal for me. Now I dream of a family, a husband, children. I need someone who'll put their arms around me and say everything's going to be all right, someone to make me feel safe. On the evening of 24 February, the day the war started, my mother wrote from Moscow to say that her pension had come in. I've got her bank

card; she uses it when she visits Kyiv. She said I must take out all the money and send it to my brother in Moscow. Not a word about the war. Not a single question about how I was doing, whether I was still alive, how I felt. I was in shock. I couldn't believe my eyes. Kyiv was under fire. But I repressed my emotions and wrote back that we were in the middle of a war, that I'd cash the money as soon as I could and try to send it. After that message she asked whether I was going to try to escape. She seemed to glorify the war, as if Ukraine was finally getting what it deserved. I didn't answer. I didn't want to talk to her at all anymore.

'After two days she sent me another new message, asking why I hadn't replied. I could no longer control myself then. I was a nervous wreck, the war had been going on for days, I couldn't go anywhere. I wrote that as my mother she could have asked whether I was alive and how I was doing instead of only talking about money. She wrote back that I hadn't shown any concern for her either! That was insulting and it hurt. I left it at that, tried not to think about her. It was more important to think about how to escape the hell of the bombardment.'

At first Margarita thought the attack would last only a few days. Throughout those early days she was on Telegram all the time, even at night, because she couldn't sleep.

'On the sixth day I decided to leave. I live in Sofia Rezidens, a new housing estate in Kyiv, close to the airport in Zhuliany. I went to Pasazhyrskyi Station with just a small suitcase and my cat Tosya. I had no idea what to do, how to live from then on. It was impossible to get on a train. After a few hours I was lucky enough to be able to get into a taxi with a mother and daughter. The driver was Azerbaijani. He took us to Uzhhorod. It was a dangerous journey, eighteen hours long. There were attacks on the railway all day. A few hours after us, the road we fled on was bombed.

'My fellow passengers weren't particularly happy with the presence of Tosya. Everyone in the car seemed to be allergic to cats, but fortunately that was okay in the end. The cat didn't do so well. Tosya couldn't handle the stress, having never travelled for more than an hour before. She lost all her fur. The trip cost me 500 euro. Without a male passenger, a car to the border cost 2000 euro; with a man in it, 3,000. The mother and daughter knew people in Uzhhorod. They took me with them, as if that was perfectly natural, as if I was family. We arrived at a house. Fifteen people came to greet us and asked what we needed. I cried.

'My Swiss ex has Georgian friends in Budapest and he rang them. That's how I ended up in Budapest. They insisted on picking me up in Záhony. The next day I went by train from the border town of Chop into Hungary, to Záhony. The moment I crossed the border in the train my mother rang. I didn't actually feel like talking to her at all, but I answered. She started crying and apologizing and saying how much she cared about me. I told her everything was fine and that I was somewhere safe. Nowadays she tries to play the part of a good mother.

'The Georgians had brought a lot of things for me and the other refugees: food, nappies, children's toys, tricycles. They even helped me to find an apartment in Budapest.

Georgians have followed me everywhere, all my life. I speak a bit of Georgian. I lived in Tbilisi for eighteen months and I was once married to a Georgian, although that wasn't a success. My ex-husband was important to me, though; he taught me to think for myself. He opened my mind. It's because of him that I started to read. But what I wanted to ask you is, I didn't bring anything with me from Kyiv, only my cat and a small bag: could you bring my clothes here from Kyiv? From my apartment?

The border at Záhony, Tuesday 24 May 2022

Staring at the sky

The fences and privet hedges around the gardens are tall. In this eastern corner of Hungary many of the houses are empty or for sale. At the roundabout next to a seed merchant's is an MOL filling station. It's the first petrol station over the border when you come out of Ukraine. It was here that Alexey stopped the delivery van in which he fled from Kyiv with his family and fell into a deep sleep lasting two hours. There's a small queue of three or four cars at each pump, Polish, Hungarian and Ukrainian license plates, and a lot of men with broad Carpathian faces.

It's three months since the war began and I drove to Záhony to pick up Alexey and Yeva and their children. Ilona had strictly forbidden me to enter Ukraine; she knows me and she wants to hang on to me for a while yet. But now it's okay. The reports are favourable. The Russians have withdrawn from the suburbs of Kyiv and they've even been pushed out of Kharkiv by the Ukrainian army, almost back into Russia. I fill the Toyota until I can see diesel churning just below the rim of the fuel tank. In Ukraine all supplies of diesel are reserved for the army and farmers. For my trip I've got eight big jerry cans on the back seat to get me from Záhony to Kyiv and back, and so I can drive around a bit too.

Next to me is the dead-straight railway embankment. I drive through abandoned villages populated by Roma women with beautiful big eyes and bad teeth, villages where only the stonemasons still have a flourishing business, engraving headstones. Jesus hangs on the cross and storks breed on top of the wooden utility poles. The bright-green leaves on the trees have transformed a border area that looked bare and sinister three months ago and made it enchanting. I drive through tunnels of creamy white blossoming acacias. Nature is sprouting wildly, as if I'm in the tropics. Nearer the border the houses are closer together. I wait ten minutes at a level crossing somewhere for a blue local train.

It's half past five and the border is still open. I'm going to Ukraine and that excites me. Aside from the fact that I've loaded up with 130 litres of diesel and an old road map of Russia, and have papers with me to explain that I work for a Ukrainian humanitarian organization, approved by the Ukrainian ministry of defence, I'm hopelessly ill-prepared for this trip. I have one appointment in Lviv and a few in Kyiv. My most important mission is to fetch Neal, the American sniper.

I've booked hotels along the way through Booking.com, as if I'm going on holiday. Initially I was planning to pick up a young woman on the outward journey, in Uzhhorod, close to the Slovak border, and take her to Kyiv, but last night I heard she'd arrived in Kyiv already. So I've lost the guide who was supposed to steer me across the country and through the roadblocks. I don't know what to expect. Someone who travelled to Kyiv a few weeks ago advised me to watch out for shell fragments on the road and take a good spare tyre, to remove my sunglasses at roadblocks, to have the right papers on me, preferably with copies in Ukrainian, and to stock up on fuel and food, because there's a shortage of both.

The protective gear we've purchased over the past few months is still being driven to the front line in delivery vans via Poland. I've asked Illya what I should take to Kyiv.

Bullet-proof vests, which everyone was crying out for in the early days of the war, are not much in demand any longer. Those extra eight kilos make you slow. Kevlar helmets are badly needed now, but almost impossible to get. Illya said that the water supply company in Chernihiv, 150 kilometres north of Kyiv, is urgently in need of tools to repair its infrastructure. He gave me a list of what they were lacking. At the OBI hardware store in Kaposvár I loaded the Toyota pickup full of angle grinders, drills, screw guns, pumps, a circular saw, extra batteries, battery chargers and chainsaws. Apart from the pumps and chainsaws, everything works on the same eighteen-volt battery, Bosch Professional. In a country where hardly any wars have been fought in a hundred years, a cordless drill is preferable to the plug-in variety, let alone in a city seventy per cent of which has been smashed to smithereens.

The queue at the border isn't long, but I'm picked out and have to unload the twenty plastic storage cases of tools and eight jerrycans of diesel onto the wooden bench provided. I'm asked to explain what I'm planning to do with the diesel. They believe me, thank god, so I'm not treated as a diesel smuggler. The rather forbidding customs men speak only Hungarian. '*Hova megy*?!' Where are you going? They open some of the packaging at random. It seems customs is serious about checking that no weapons get into Ukraine from Hungary. When the bespectacled boss sees that I have a bottle of palinka with me and I tell him it's the family palinka, which we distil in the village in Somogy, at the local distillery, he relaxes and I'm waved through.

I cross the River Tisza on a narrow steel bridge and drive into Ukraine. It immediately feels different, as if even the birds are quieter here. At the end of the bridge, two young, English-speaking soldiers are waiting for me in camouflage uniforms, armed with AK-47s. On the Ukrainian side of the border I'm scrutinized at length and sent

from hatch to hatch with messy paperwork that needs to be stamped, as if in a relay race, in the old Soviet style. There are a lot of soldiers everywhere. The organization whose papers I'm carrying is checked out particularly carefully and I'm interrogated as to whether I might be planning to leave the pickup behind – all in a congenial but thorough manner. My things are only superficially examined and I don't have to open any bags or plastic storage cases.

Then, under a wide sky, I drive into a country at war. A vast, flat, empty landscape begins immediately after the border. Not a soul to be seen. Except for a Roma man with a guitar on his back riding an old bicycle, on his way between two villages, like a shot from a Kusturica film. The filling stations are deserted. A railway building has white sandbags piled up in front of its doors and windows. To the left and right is a lot of unworked, overgrown farmland; to judge by the height of the bushes and trees it hasn't been ploughed for about fifteen years, gradually neglected after the collective farms were broken up. The heads of goats and rams stick up above the reeds. The villages look the same as on the Hungarian side: linear settlements. In the evening sun, elderly women lean against the wooden fences, many dressed in black, occasionally accompanied by an old man.

They watch me pass. On the paths of red dirt leading off from the road, gaggles of geese pad about. The only difference is that in the villages there are a lot of cyclists, on beautiful old Warsaw Pact bicycles. It's only later that I realize this has nothing to do with poverty or a preference for cycling but with the lack of diesel.

I'm on the qui-vive and I catch myself scanning the sky for incoming rockets. Completely ridiculous, of course; I'm a novice, it's the first time in my life that I've travelled through a country involved in a serious war. I feel the excitement and the immense space around me. In the distance are the slopes of the Low Carpathians, blue. A Roma man comes towards me along the hard shoulder, riding a skinny horse, galloping, without a saddle, without stirrups, jolting on its bony back.

Mukachevo, Tuesday evening, 24 May 2022

Destination unknown

The promenade in Mukachevo is the way promenades are meant to be, forty metres wide and lined with horse chestnut trees on either side. A century ago this street was still called simply Horthy Miklós út, the central square was Árpád tér, next to the hotel was Farkas ruha (Hungarian for the Wolf family's clothing shop) and the city was called Munkács. I have a short journey ahead and I've chosen my hotels with some care. My first stop on the way to Kyiv is in Transcarpathia. My thinking before I left was as follows: Putin *bácsi* (Uncle Putin) will not want to lose one of the last friends he has in the world by bombing an agricultural region that's home to a relatively large number of ethnic Hungarians.

On inquiry it turned out that only five per cent of the residents of Mukachevo are ethnic Hungarians, even though this western region of Ukraine was part of the Kingdom of Hungary for a long time. On display in the corridor of my hotel are black-and-white photos from the early twentieth century, showing a charming little country town. Not much has changed apart from the street names; it's a place typical of the Dual Monarchy, with a haze of Habsburg *Jugendstil* blanketing everything, its broad streets, its pretty, frumpy houses and imposing public buildings, its post office, theatre and town hall – all built to stress the solidity of the empire. The social stratification of a small town like this a hundred years ago, and its provincial splendour, were captured inimitably by Dezső Kosztolányi in his novel *Skylark*, about a couple who worry that their only daughter will never find a husband. The town in the book was modelled on Szabadka, but it might just as well have been Munkács.

I'm surprised by the number of able-bodied men on the streets. I assume that if you can show you're indispensable, as a doctor, firefighter or farmer, you don't need to go to the front. I'll have to investigate some time and find out what circumstances qualify you to stay at home with your wife or mother, warm under the covers. The hotel car park is full of white vehicles belonging to the United Nations (you wonder why they need quite so many pickup trucks) and the entrance to the police station is barricaded by a high wall of white sandbags with gaps on either side to shoot through. The receptionist has warned me about the curfew, saying there are hardly any people on the streets then, and a lot of police, who will stop you and ask what you're doing.

I go looking for a place to eat and find a restaurant called Villa del Re. I have to take the lift to the third floor, a lift full of gleaming gold and mirrors, with marble on the floor. The lift doors open to reveal a pre-war Ukraine: a lounge bar with mud-brown and seaweed-green chesterfield sofas, a cocktail bar with rows of sparkling bottles and pounding music. A glass extension offers a view out over the city. The guests are all young women with fake designer handbags, imitation haute couture and pimpled cheeks. They're still serving. A young man comes up to me. He speaks good English and explains with an absurd degree of kindness – he seems to regret it from the depths of his heart – that unfortunately there is no food left.

'The kitchen shuts at nine o'clock, given the situation.'
'Do you have a gin and tonic?'

'No, sorry, strong drink is served only till six in the evening, in view of the difficult situation.'

I take a walk through the centre of Mukachevo, past four restaurants, and feel as if I'm in *Le Charme discret de la bourgeoisie*, a Buñuel film about two couples looking for a restaurant. Everywhere I go the kitchen is shut. Fortunately I come upon a supermarket that's still open. It's getting dark. A little later I'm walking down the broad promenade of Mukachevo with a paper shopping bag when the air-raid siren goes off. I'm told that when that happens, a cruise missile has left Russia, exact destination unknown. No one reacts any longer, since the target is normally somewhere else. Parents let their toddlers run around while it sounds; people don't even look up from their phones.

In my hotel room is a big television, and while I'm undressing and brushing my teeth I watch energetic promotional films for the army: abseiling paratroopers, low-flying helicopters, troops storming ahead. They're interspersed with music videos by Ukrainian bands that have thrown themselves into the war: sunglasses and guitars amid wide fields of rape or sunflowers, a blue sky above, along with everything else that's beautiful about Ukraine. The music videos also feature a lot of heavily armed soldiers, children singing, overwhelming armoured attacks and the prettiest of women in folk costume. It's all very slick. The highly nationalistic, war-glorifying kitsch makes me nauseous.

When I speak to Illya a few days later and say that I find it repulsive television, adding that although I understand you want to control the news coverage completely during a war, it could be toned down a bit, he tells me that for a limited period all the broadcasters have been merged and there's just one Ukrainian channel, this one. He agrees that it's hideous. The reporters who normally work for channels two or three continually announce

themselves as being from this one broadcaster, which is a ridiculous ritual. 'It's crucial that all the diverse voices and critical debates come back immediately as soon as there's peace in Ukraine,' Illya says.

From Mukachevo to Lviv, Wednesday 25 May 2022

Dense forests

The centre of Mukachevo presents a convivial street scene. The sun is shining, the pavements and café terraces are packed, the war seems forgotten. Everything suggests that the Ukrainians are *bon vivants*. They don't like getting up early, preferring the Mediterranean mentality to the German, in fact they're night owls. This is based on nothing more than extrapolation after half a day in the country, incidentally, and on observing the handful of Ukrainians I happen to have got to know over the past few weeks.

I write the names of the places I need to pass through on my left hand and try to memorize how they're written in Cyrillic script. In this region there are occasional road signs. I drive through woodland from Mukachevo to Lviv. All the bridges are guarded by soldiers behind sandbags, who have taken up position on all four corners. Under the foliage in the woods close to the bridge there will often be a larger and heavier structure made of tree trunks, with yet more troops. In the mountains near Dubyna I buy a wolf pelt complete with head. Ever since I read *Never Cry Wolf* by Farley Mowat as a boy, I've loved the animal and keep hoping I might happen upon one during my wanderings through the forests of Eastern Europe. The enthusiasm for this large carnivore in the manicured park known as the Netherlands strikes me as odd. Wolves are social animals that usually live in packs, with hunting grounds of between 50,000 and 200,000 hectares. There are still plenty of stretches of natural landscape of that size in Eastern Europe, but not in the Netherlands any longer.

I asked Neal whether he'd seen many wild animals during his days and nights in the trenches. He'd told me he always trained soldiers on location and stayed with them in the trenches for a few days, at places where there was no fighting at that point. He hadn't seen any wolves or bears, but there were red foxes, even at the training centre in the middle of Kyiv. The city fox was fearless despite being surrounded by armed men. The troops fed the animal, as a kind of local mascot. Neal kept calling it 'little fox'. The foxes at home must be rather more substantial. On WhatsApp he sent me a photo of a wild boar in Alabama, a monster, as big as a young buffalo.

In the fields around the trenches, cranes walk about in groups of twenty or thirty. They nest locally. I thought cranes bred further north, up in the Baltic states, but it seems they also populate the wetlands of Zaporizhzhia, where Neal had been living in the trenches for several weeks with the Ukrainian forces. He told me that every unit has at least one cat, sometimes several. The cats are well cared for and eat the same food as the men. Mice and rats are followed by snakes, and both the common adder and its cousin the steppe viper have a nasty bite. The platoon cat keeps the trenches free of both rats and snakes.

Every few nights he would return to the barracks in an old factory near Zaporizhzhia, where he saw his friend the gunsmith. From there he went from unit to unit to convey the finer points of warfare to eager dentists, teachers and factory workers. The semicircle of reinforcements on the banks of the Dnipro at Zaporizhzhia is an important defensive line for the Ukrainians, Neal tells me. The area in and around the trenches and underground bunkers teems with squirrels, and out on the fields are 'little deer'. They must be roe deer, a slender species unknown in America, a land

where everything is big, even the trees. The Americans have embraced Oliver B. Bumble's subtle notion of beauty: 'Big is beautiful, lots is lovely.'

The mountains I'm winding through are a little lower than the Carpathians at Brașov in Transylvania. In the distance just a few are topped with snow. Villages lie hidden in the valleys. In the shade at the edge of the forest, Roma are selling berries, mushrooms and bottles of fruit juice. The pass to the Sambir district is an unmanned fortress with walls of sandbags, next to one of which stands a life-sized gleaming replica of an American deer with a white bib. Beyond the pass the vegetation suddenly changes and mixed deciduous woodland gives way to pine forest, with the occasional light-brown, bare stretch where the bark beetle has destroyed a plantation of spruce. The people here are spared nothing.

The villages follow the rhythm of spring, Putin or no Putin. It's getting milder and more pleasant. This is an undulating Alpine landscape with beautiful old wooden houses, pink and white blossom in the villages, the scent of wood fires, and women at the roadside with sweet peppers and tomatoes for sale. Every few villages there's a checkpoint with soldiers chewing sunflower seeds. The improvised bunkers cling to the hills and slopes like fresh wasp nests.

I pass filling stations where the pumps, including the hoses, are wrapped in transparent cellophane, rather like the way that in the first few days of the war people tied informants and looters to lamp posts, their trousers pulled down and their buttocks bare, so that babushkas could thrash the traitors with willow switches. In early March I regularly saw films on Telegram of tied-up informants, but not any longer. The photos and videos in the groups I belong to on Telegram have become a bit milder, more bloodless. Traitors will undoubtedly still be punished, but the Ukrainian ministry of defence has got a grip on the internet, so that on Telegram I now see mainly statistics about destroyed tanks and dead Russian soldiers, women with flowers in their hair, or children singing patriotic songs. There are fewer rocket attacks on Ukrainians cities, in fact there are probably fewer attacks altogether at the moment. It was gruesome but hypnotizing. The ball of fire moving across the sky, the bang and the flash. And, when a bombardment ended, the silence, with just the lonely wail of car alarms.

Lviv, Wednesday evening, 25 May 2022

Hi baby, love you

In my own way I've done my best to make clever hotel reservations on Booking.com, by which I mean safe. No highrise. No big international hotels. In Kyiv I'll be close to an old Orthodox church and opposite the botanical gardens. In retrospect I realize the effort was misplaced. Putin couldn't care less; Orthodox churches are being flattened out of spite. The complete destruction of everything that's Ukrainian is the aim. In Lviv my attempt at safe booking has failed completely. There's nothing but glass around me. The hotel is a stone's throw from the railway station and it seems to be filled entirely by foreigners who have come to make Putin's life a misery.

Next to the lift doors on each floor is a plan showing how to get to the bomb shelter. The air-raid siren wails, its sound blowing in through the open windows. Nobody pays any attention and I adjust to local customs. I sit in the restaurant of the fashionable hotel with a view of the city, an aquarium on the sixth floor, surrounded by men every single one of whom looks as if he's stepped out of the thriller series *Fauda*: fit, muscular, shaven-headed, sporting tattoos, thick beards, hard faces, practical clothing, running shoes and a feline walk. Groups of men are consulting in English and Ukrainian. I catch exhilarating snatches: 'We have one day to get in, one day to get out.'

Next to me someone says to those around him, shaking his head to indicate that it must never happen again, 'I was on my own that time.' The air-raid siren keeps going off, which is annoying mainly because the kitchen closes in response and you can get only a rather unappetizing 'mixed plate': chunks of cheese, slices of salami and dry white bread. They're just lazy, I suspect, the cooks here, exploiting the air-raid warning as an excuse to go home early.

Behind me an American serviceman has his phone on loudspeaker. A woman is carping at him in lilting English. Whether he's drunk, she wants to know. He's a giant of a lad with a thuggish face, with whom you could no doubt enjoy a good night's drinking, but to judge by first impressions not someone I'd stick my hand in the fire for when it comes to marital loyalty. He does have an excellent tactic, one we could all learn from. He's very patient, carries on drinking his whisky-on-ice, ignores the nasal bleating, doesn't address any of the points made, rotates the ice cubes in the heavy glass and repeats his mantra. Dozens of times. Every twenty seconds or so he says it: 'Hi baby, love you.'

Asking the way

At the hotel in Lviv there are two ambulances in the car park with a small gathering of older German-speaking gentlemen around them. Just as the resistance to Russian violence in Ukraine came from the people, from the bottom up, so foreign aid was started on the initiative of ordinary citizens, long before the European Union or countries like Germany and France made a move.

I walk through the city centre. Elderly women lugging plastic bags are overtaken by boys slaloming on fluorescent green scooters. Some of the windows have diagonal crosses stuck on them in white tape, to restrain flying glass in case of bombardment. One detail that strikes me: people walk quickly, although not as hurriedly as I always walk. In Lviv there's none of the lethargic sauntering that normally drives me crazy in shopping streets. People stride along nicely. No time to lose; tomorrow you might be dead. Furthermore, people in Lviv care about their appearance. I pass a 'Barbershop', a 'Face Factory', a 'Beauty Hall', a 'Hair Factory' and a 'Nail Factory', all close together, all working flat out.

At the town hall a Ukrainian poses in a red floral folkloric dress, wearing a tiara of sunflowers in her long blonde hair. She's being photographed. Possibly a bride. Lots of people get married in wartime. You see weddings and brides everywhere, both on Telegram and in real life. I don't know whether there are pensions for widows and orphans left behind. On the walls of Lviv, paintings in blue and yellow glorify self-sacrifice. In one a woman ardently takes leave of her lover. They are the same images as were used a century ago.

In our house in the Hungarian countryside I have several pictures from the First World War. In those days too, there was a need to inspire enthusiasm. It's an integral part of warfare. In one picture a hussar is saying farewell to his sweetheart and his dog. A raspberry-red cape with gold braid hangs over his shoulder. A short distance away, his comrade is waiting with his saddled horse. In the next picture he's embracing her with his arm in a sling, having returned from the front, and in the background the other hussars watch from their mounts. You hope that all the young lads will arrive home as lightly wounded as this particular hussar.

War intensifies masculinity and femininity. In the streetscape I see a lot of men walking with resolute strides, looking tough in their green hats, shirts and trousers, including the non-soldiers – the visible militarization of a society. Men in uniform move along the pavements at marching pace, their arms swinging back and forth at their sides, on their way to the bus station to return to their units.

At a café on an intimate square lined with nineteenth-century houses I drink coffee with a computer programmer from Odesa. He tells me everything's fine there, aside from the occasional bombardment, and that he likes the author

Isaac Babel because he writes so well about Odesa. He says his grandparents are Russian, and that Putin has moved people to Donetsk and the Crimea from all over Russia, but that nevertheless, still only ten per cent of people in Ukraine are pro-Russian, if that. 'Everyone hates the Russians.'

Since I've seen so many men of call-up age walking about in civilian clothes, I ask him how mobilization actually works.

'In the first wave of mobilization all men with combat experience had to report for duty. In the second wave everyone with military training will have to sign up. That second wave of mobilization is expected to happen soon. In the third wave, all fit men of fighting age will need to report for duty, and the fourth wave will be for every living Ukrainian man, without exception. Right now you can avoid being recruited if you're contributing in some other way, as a volunteer, or if you can show that you're important for the economy. That will no longer apply in the fourth wave. Everybody will have to fight then.'

I drive out of Lviv, that beautiful Habsburg city that you immediately feel you want to live in, a city that exudes an atmosphere of youthful mischief with its dilapidat-

ed houses, a city with cast-iron balconies and weeds growing out of Jugendstil facades. Today the sun won't help me by showing me the way to the east. It's cloudy. I hate asking for directions and I don't have Google Maps or Waze. After Lviv all the road signs have been removed, including the street names in every town and village. Tarpaulins or agricultural plastic have been stretched over the grotesque Soviet artworks on the edge of big cities and at provincial boundaries, since they include the names of places and provinces; sometimes the names have been obliterated with a hammer and chisel. The scale of the road map I have on me is one to two million. I bought the map a quarter of a century ago when I drove from Kyiv to Odesa to see the Isaac Babel exhibits in the Literature Museum.

On 15 May 1939 the NKVD arrested Isaac Babel at his dacha outside Moscow. Babel was a writer, a Jew and a native of Odesa, any one of which was sufficient reason to pull a person's fingernails out and then put them up in front of a firing squad – in those days in Moscow and again now. But in addition, as I understand it, Babel had been making advances to the mistress of one of the leaders of the NKVD. Now that strikes me as serious daredevilry.

'You have been arrested for treason and anti-Soviet activities. Do you admit your guilt?'

'No,' Babel said. 'I don't.'

'So how can you reconcile this declaration of innocence with the fact that you've been arrested?'

In the Literature Museum in Odesa I saw his fragile spectacles in a glass case. He was tortured in the notorious Lubyanka prison and signed a confession after three days. He was executed a few months later at one-thirty in the morning and then tossed into a mass grave. His final words were, 'I am asking for only one thing – let me finish my work.'

Now that I'm travelling through Ukraine, I've taken out of my bookcases works by people who have written about this part of the world. They are among the best writers I know, and this new war has brought them together in my travel bag: Isaac Babel (*The Red Cavalry*), Joseph Roth (*Reisen in die Ukraine und nach Russland*), Curzio Malaparte (*The Volga Rises in Europe*) and Gregor von Rezzori (*The Snows of Yesteryear*).

The map I have with me is the Freytag & Berndt road map of Russia and the Commonwealth of Independent States, which includes Ukraine. Even the oil pipeline is on it, in thin grey lines. On the map I see place names that I recognize from *The Red Cavalry*: Novohrad-Volynskyi, Brody, Kremenets, Berestetshko. It's a thrill to drive through this landscape, through towns and villages where Babel advanced with the Cossacks against the Polish magnates and wrote his inimitable stories.

The land is bare, a pale brown. Many of the fields have not been cultivated. One difficulty is that after Lviv the landscape is open – fields, meadows, few woods near the road – so I can't see on which side of the tree trunks moss is growing. Those are the only two aids to orientation that I have, sun and moss. One is to the south, the other grows to the north. In essence it's simple: if I drive directly eastwards, I'll arrive in Kyiv as a matter of course.

Galicia, Thursday 26 May 2022

Black gold

Four times over, I had to ask the way. 'Kyiv,' I called out, and people pointed in the right direction. I passed columns of ambulances and fire engines with German, Swiss and Swedish license plates. When I first found myself following three fire engines with blue flashing lights between Lviv and Busk, I thought there was a fire, or that a bomb had hit, but after a while it dawned on my slow brain that they were donated vehicles, on their way to disaster areas in the east.

During the last ice age, a layer of humus a metre and a half deep slid over the land, covering two thirds of the surface of Ukraine. Chernozem consists of very fertile black soil, rich in humus and oxygen, with a high water-storage capacity. Ukraine has a quarter of all the world's chernozem (a word derived from *chorny*, 'black' in Russian, and *zemlya*, 'earth'). It was the reason why Hitler wanted to conquer Ukraine, to be able to feed his troops and his people. It was also the reason why everyone was always eager to annex this land, and why the Polish magnates kept the serfs here by force to sow and harvest.

The chernozem, or rather the grain that grew on it without any need to apply expensive fertilizer, made many people a fortune, from the Polish magnates to the major Jewish grain dealers in Odesa, but not the Ukrainian serfs who worked the land. Those who fled that harsh life formed groups of Cossacks. Their lives remained hard, but they were free men. The Cossack community with its Orthodox

faith grew, until at one point there were twenty official regiments, each of two thousand Cossacks. Between 1649 and 1764 the Cossacks formed their own state, the Hetmanate, with a hetman as head of state.

Thirty kilometres short of Brody, on a lonely hill in a marshy area, stands Olesko Castle. Several centuries ago, this was the home of the man who later became king of Poland as John III Sobieski. It was he who broke the siege of Vienna in 1683 with a downhill cavalry charge, putting to flight the Ottoman army under the leadership of Kara Mustafa Pasha, which included 40,000 Crimean Tatars on fast horses. I turn off the E40 and take a narrow asphalt road that's overgrown by shrubbery.

At the entrance to Olesko a thick grey tree trunk bars the way. On and around it are white bags filled with fine sand. Behind a bush is a hut, four by four metres square. There's no one around. The wind howls in from far away; above me is the six-hundred-year-old Olesko Castle and ahead of me three-month-old defensive works. The hut is covered with agricultural plastic. In the corner is a simple bed, just a few planks laid over beams and covered with straw. Next to it is a rusty round iron stove and a pile of firewood. It moves me, this boys' den. The defensive works, just one example of hundreds I'll come upon, remind me of the structures my sons made with their friends near the old coffin factory in Hungary. Next to this one, car and tractor tyres are stacked up, and two of the stacks have long branches poking out, flying the Ukrainian flag: yellow for the grain, blue for the sky.

I park the Toyota, buy a few things at the tourist stall – you have to help keep the economy going, after all, even in Olesko where not a soul comes – and walk up to the fortress. In the courtyard is a chubby security guard with a pistol in the holster on his hip, the only male amid dozens of dowdy women in dust jackets. In the entrance hall three of them are waiting to check my ticket, and there are two in almost every room as attendants. I'm the only visitor. I look at paintings from John III Sobieski's collection and am amazed by the way some of the Polish magnates, with their turbans and brightly coloured robes, look just like Ottomans. They're indistinguishable from members of the court of the sultan in portraits by Jean Baptiste Vanmour. One huge painting shows the Polish hussars under Sobieski's leadership storming down the Kahlenberg and laying into the Turks, who flee in all directions in horror. I stare up at it for several minutes, even though I really ought to be hurrying to get to Kyiv before the curfew.

Kyiv, Hotel Hermitage, Friday 27 May 2022

War flag

My senses are in a heightened state of readiness. Which I do quite like. All my impressions are more intense, because I'm alone, and in a country at war. I don't know what to expect. I'm quickly getting accustomed to many aspects of the war: the lack of diesel, the abandoned filling stations, the checkpoints, the many soldiers everywhere, the blocks of concrete on the roads, the empty streets, the sandbags in front of cellar windows, the crosses taped on glass, the air-raid siren, the curfew, the closed restaurants and cafés.

In the region between Lviv and Kyiv, almost every village has raised not just the national flag but Ukraine's war flag too, red and black, as a sign that they're ready to fight to the death. It's pirate-like. Black for the earth, red for the blood that will inevitably flow. Not many countries have a war flag, a special flag for their armed forces. Germany, Britain, Poland and Israel have one, as does Ukraine. The flying of the war flag is a custom of the extreme nationalists in the area around Lviv.

The fact that every village has paid for its own fort on the road leading into it shows the degree to which resistance to the Russians has come from below. The farmer's sons who have used their tractors to drag blocks of concrete are making Putin's war difficult, expensive and impossible to win without scorched-earth tactics. Taking charge together, rather than creeping away into a corner like frightened birds, gives power and cohesion to the community. It's this strength and cohesion that Putin didn't expect, to judge by the dress uniforms and riot police shields and helmets found in the columns of Russian army trucks abandoned between Chernihiv and Kyiv.

On the hills I see trench systems, and there are dozens of roadblocks and hundreds of improvised bunkers abandoned at the side of the road to Kyiv after the Russians withdrew from the north of the country. I stop a few times, including to piss, and walk into the trenches, rather hurriedly and timidly, fearing the local militia might mistake me, with my Hungarian license plates, for a Russian saboteur and impale me on a pitchfork.

Around eighty kilometres from Lviv the road widens from two lanes to four, with a crash barrier in the middle. I can accelerate. Except that someone has had the bright idea of painting zebra crossings on the highway, quite a few of them too, so that the locals can saunter across, which happens from time to time: mothers with children, bandy-legged farmers, sometimes wheeling a bicycle or leading a cow on a rope. Horse-drawn carts use the hard shoulder, while occasionally BMWs and Maseratis pass me in the left-hand lane at 180 kilometres an hour. Between Novohrad-Volynskyi and Zhytomyr I start to doze off. If I want to avoid driving in the dark, I'll have to make haste.

At 150 kilometres short of Kyiv I'm fortunate enough to be overtaken by two pickup trucks in army green that are going flat out. I decide to follow them. My green Toyota Hilux matches them to perfection. They drive aggressively, forcing everyone else off the road, often close behind each other. The E40 remains four-lane with zebra crossings. At 160 kilometre per hour we make our way eastwards. The driving speed keeps me awake.

On the outskirts of Kyiv we pass burned-out filling stations, shot-up houses and the occasional blown-out Russian tank on the hard shoulder. None of it makes them brake. At a certain point they leave the road and stop at a completely blackened filling station. Everything for a hundred metres around me is black. I get out, walk over to the driver of the rear car – he's in civilian clothing – and explain that I'm on my way to Kyiv without GPS and ask where they're going. They're en route to the front in Donetsk to deliver the two vehicles and need to hurry, but they'll take the time to lead me to my hotel.

We hurtle along dikes beside collapsed bridges, through suburbs demolished by shelling. Soldiers are digging trenches at the edge of Kyiv. At the city boundary, a heavily fortified checkpoint is manned by dozens of troops with Kalashnikovs. In Kyiv we barely slow down, taking Victory Avenue into the centre of town at 140 kilometres per hour. My guides stop in front of Hotel Hermitage. I thank them and they tear off.

As soon as I walk into my hotel room, the air-raid siren starts up. I go over to the balcony and scan the sky. The hotel is an old building, three floors high. I'm on the first floor. I open the windows and close the curtains. Assuming that the kitchens in Kyiv shut when the alarms sound, as they do in Lviv, I get into bed accompanied by Babel and Malaparte. There's a curfew from ten in the evening until six in the morning – the times vary according to the situation and the city or town. The curfew is actually rather pleasant. It liberates you from the gnawing feeling that you're missing something.

The city is quiet. The streets are empty, apart from a few taxis. Then a downpour begins, and everywhere water leaks onto metal structures in the maze of parking places, cobbled together garages and balcony roofs in the courtyard behind the hotel. It sounds like a gigantic washing machine that's just about to break down.

Kyiv, Café-Restaurant Gorchitsa, Friday evening, 27 May 2022

Don't worry, they only bomb the Nazis

'On 24 February we had a photo session planned for the launch of our new CD,' Anastasia tells me. 'My mother rang and woke me up at four in the morning. She lives in the western part of Kyiv, over towards Irpin, and she'd heard explosions.' Anastasia is a singer, twenty-eight years old, with a beautiful face and long black hair. She's shy. She's sitting diagonally across from me in Café-Restaurant Gorchitsa, close to the heavily guarded Ukrainian government quarter, and she barely looks at me. Next to me sits her boyfriend Sasha, a tall, quiet young man.

I'm fifteen minutes late for our meeting because I wasn't allowed down many of the streets and kept being stopped and checked by heavily armed soldiers. The streets in this district are closed off with blocks of concrete. Men wearing bullet-proof vests sit in bunkers made of concrete slabs and sandbags, and a short distance away are armoured vehicles. The government quarter is a fort. The streets are empty and it's pitch dark. The restaurant, where we are the only diners, belongs to a Frenchman who has lived in Kyiv for thirty years. He came to fetch me from where I stood with a soldier, who wouldn't let me go any further even after I'd shown him my passport, twenty metres from the door to the restaurant. For the first time in Ukraine I'm having a warm meal – for three evenings in a row an air-raid cheated me of that. The food is excellent.

'My mother's boyfriend was in hospital,' Anastasia says. 'We fetched clothes and took him away from there. When the air-raid siren went off we went into the shelter with my mother. She has difficulty walking. The next day there were a lot more explosions. It was getting closer. We got my mother and her boyfriend to a place of safety that day. My father lives in Russia. He's Russian. I was on the phone to him during an air raid, on that second day of the war, and he said, "Don't worry, they won't bomb you; they only bomb the Nazis." He told me that if Russian soldiers came to the door, I must tell them that both my grandmothers are Russian; that meant I was de-Nazified. I told him to go to hell. I haven't spoken to him since then.'

Anastasia is tense. As she speaks she looks out of the window, or at Sasha, or at the tabletop.

'My father was a colonel in the Ukrainian army. Immediately after the Maidan revolution he went back to Russia,' she says. 'He always screamed and swore at my mother, and he hit me. They divorced when I was nine. He worked from eight till six; I don't know where he was or exactly what he got up to all day. He has two sons and a wife in Russia. I hate him.'

Sasha is a musical genius who plays dozens of instruments. His mother comes from Okhtyrka, between Sumy, the provincial capital, and Kharkiv, close to the Russian border.

'They refused to leave Okhtyrka, my mother and her sister. No matter what I said,' Sasha tells me. 'I begged them, but they stubbornly insisted on staying. In the end I said I was going to wipe their numbers from my phone and added, "And I'm going to start forgetting you right now!" That worked. They came by train. It took twenty-five hours – the journey normally takes four hours – and on some stretches the train went no faster than ten kilometres an hour, or had to stop in the middle of farmland because of the fighting. The train was blacked out and everyone had to turn off their phone so they wouldn't be targeted.'

'We took both our mothers to safety and then started to think about what to do ourselves,' says Anastasia. 'Our flat's on the Irpin side – there's no way we could stay. There was heavy fighting in Irpin. We went to our practice and recording studio to put our things somewhere safe. When we got there it occurred to us that the sound studio was the perfect place to shelter. It's underground. We furnished it and told friends they could come too.'

Kyiv, underground studio More Music, Friday 27 May 2022, a little before midnight.

We sometimes forgot the war was so close

At eleven in the evening we leave the centre of Kyiv, taking a taxi from Pylypa Orlyka Street, close to the ministry of culture, along increasingly narrow roads. We drive down Yuria Ilyenka Street at full speed, swing through the bend, then turn right at Ovrutska Street before winding down a steep hill lined with trees. Along a bumpy road we arrive at a cul-de-sac in a small, nondescript trading estate: 12 Delehatskyi Lane. I follow Anastasia and Sasha out of the taxi. They open a large metal door, beyond which it's pitch dark. Anastasia turns to point to a sign on the wall that I can just about read in the glow of a streetlamp.

'Look, here it is, our studio. More Music. There's a plastics factory nearby that belongs to friends, over there. They do humanitarian work too. And there's a practice studio, right next door.'

'We've lived here with friends ever since the war started, from the end of February,' Sasha says. 'Along with a dog, a cat, a rat and a squirrel. There were a whole lot of cats and dogs here in the courtyard that we cared for as well. From time to time someone would go up to fetch food. Tonight's the last night. The lease has ended. The owner wants to do something else with the place.'

Behind the steel door, steep steps lead down. I follow Sasha and Anastasia, as if descending into the engine room of a ship. On the walls little scenes have been painted, with care and by a good painter: landscapes, buildings.

'All the others have left over the past few days,' Anastasia says. 'There used to be a big banner on this wall saying "Russian warship, go fuck yourself."'

We arrive in a hallway with an L-shaped bar. On the wall is a cartoon-style painting showing two musicians playing wildly, one with a double bass, the other a violin. On the bar is a microwave with cups and plastic storage containers beside it, and on a shelf above it are piles of books. The cellar has a high ceiling. It's as messy as a student flat.

'We dragged mattresses and blankets down here and invited friends. Eventually there were seventeen of us most of the time. In that space over there' – Anastasia points to the corner behind the bar – 'two friends slept with their dog. The paintings on the walls are by friends who make music and paint.'

From the moment we came through the big metal door and down a long flight of steps with a turn midway, I've been reminded of the film *Underground* by Emir Kusturica, about a group of friends who go into hiding in Yugoslavia during the Second World War. Seventeen musicians and artists in an underground recording studio in Kyiv – it's insane. I ask whether they made a lot of music together in those three months.

Anastasia shakes her head. 'I couldn't sing; I couldn't even eat or drink. I had my phone right in front of my nose all the time. I followed the news and my messages every minute.'

'At first I couldn't make any music at all,' says Sasha. 'The war, the shock, the horrors that took place: it was impossible to play. But after a month I started playing again. At first for just a few minutes, after that more and more each day.'

He opens a door and we arrive in the practice and recording studio. There are musical instruments in holders everywhere, like rifles in gun racks, ready to be pulled out, as well as three microphones on stands, and speakers the size of washing machines on either side of the drumkit.

'This was the coldest place in February and March. Two people slept here, and two or three on that side,' says Anastasia, pointing around. 'There were always between four and six people sleeping in the studio. And the cat, a giant marmalade cat.

'So there were four pets with us in the shelter. The tame rat died in April. Its owner said they were the best weeks of its life. The rat had never been given so much attention; it was very happy here. Then there was Lola, a Jack Russell. Lola is the sister of Patron, Ukraine's most famous sniffer dog, who has detected hundreds of mines. President Zelensky awarded Patron the Order of Courage for his work.'

The walls of the recording studio are made of ribbed material, a kind of rubber or plastic.

'You can't hear anything in here. It's very well insulated. We didn't hear the air-raid siren, or the bombardments, or the explosions. That was dangerous, because sometimes we entirely forgot the war was so close,' Anastasia says. 'One night we went up to smoke cigarettes. The moment we got outside the bombing started. There were explosions nearby. There are a number of military targets in this area. I saw the buildings moving around me and had a panic attack.

'In March it was freezing cold. We made a door here; these were originally two separate studios.' Anastasia proudly points to the place where they hacked a doorway out of the thick wall. 'That meant the two spaces were linked and we could use the toilet without having to go outside, and we had two flights of steps, which was an advantage during a bombardment, having two ways out. And look, here's our room!'

It's a tiny space, two metres square, the floor filled by two mattresses.

'I call this my "death chamber". I'm going to my death chamber, I'd say. I'm going to die. To be alone for five minutes, without consultation, without conversation, without organization, without a phone.'

Anastasia touches the wall. She strokes it gently with her fingertips.

'This studio has existed for twelve years and now it's going to disappear,' says Sasha. 'I worked here for six years, from 2016, as a sound engineer and producer. We've lived here for the past three months with Vira Brenner of the folk metal band FRAM and with members of the folk rock bands Krasna and O.Torvald.'

We walk back to the big sound studio. Sasha goes and stands at the piano and starts tinkling. Anastasia gets out a duduk, a double-reed woodwind instrument that comes from Armenia, and joins in. Together they play a melancholy tune. She slowly rocks her head back and forth and looks at the wall.

Kyiv, underground studio More Music, Friday night, 27 May 2022

Lifesaving microwave

'From this underground studio we did humanitarian work,' Anastasia tells me. We sheltered in here and helped people. Elderly people, children, soldiers. I was mainly searching for medicines. There was a huge shortage of those. I first tried to find them in Kyiv, then elsewhere in Ukraine, Poland, online. There are a lot of old people without their prescribed pills, and children in hospitals without the medication they need. Insulin for diabetics, for example, or some hormones, and treatments for cancer. There was nothing more to be found in Kyiv; there were long queues at the pharmacies.

'We also took food to people in Irpin, Hostomel and Bucha. Right at the start, Sasha and I took loaves of bread to Irpin and Bucha. We drove fast and kept moving, before the Russians discovered us, and threw loaves left and right out of the car windows for people, who had to grab them quickly. The Russians were shooting at everything and everyone. Later only Aleksandr delivered food. I didn't go to Irpin again until the Russians withdrew, shortly after it was liberated.

'One of the lads who sheltered in here with us, Aleksandr, an astronomer, delivered food on an electric scooter. At one point the bridge over the River Irpin was blown up. He would go to the bridge on the scooter with a big rucksack. He could then pick his way across, and on the other side he got onto a bicycle that he'd hidden among the ruins of the bridge. By bicycle he delivered supplies for those in need. He left the food at agreed places, where it was collected by the people of Irpin and Bucha.'

What a hero, that spaceman, cycling into the hell of the Kyiv suburbs. It's been estimated that around a thousand residents of Bucha were killed by Russian soldiers, and several hundred in Irpin. In Bucha the Russians made a walled children's camp their headquarters and took prisoners there. Many of them were tortured, and how many women were violated in Bucha and Irpin we'll probably never know. The Russians kept several women imprisoned in a cellar in Bucha. Nine came out alive and pregnant. The water and electricity supplies failed, so people had to leave their houses in search of food and water. Once outside they were often shot at or captured by the Russians. In Irpin the Russians drove tanks over the bodies of civilians they'd killed.

We go back into the hallway near the entrance to the recording studio, where the metal steps lead upwards. Anastasia taps the shabby-looking microwave on the bar.

'This is a lifesaving microwave. Aleksandr is from Irpin. His house in Irpin was blown up by the Russians and almost everything was destroyed, but the microwave was fine. It struck him as useful, so he pulled it out of his collapsed house. To protect the glass dish in the bottom he stuffed a cushion inside. On the way here he was shot at by Russian soldiers. He dived to the ground and took cover with some Ukrainian troops. They lay there waiting for two hours. Aleksandr was holding the microwave in front of him. When he tried to move on, a sniper fired at him. The bullet was stopped by the microwave. Look, you can see.' Anastasia shows me the bullet hole.

She stares at the floor. With her long hair and dark eyes she looks like a Native American. She's haunted by the fate of all the vulnerable people. Before the invasion began on 24 February, she used to organize camps for children from the east. Many of them are now in Russian-occupied territory. She hasn't heard from some of them for months. Those children and what she saw in Bucha keep her awake at night.

Kyiv, Hotel Hermitage, Saturday morning, 28 May 2022

How to dress for a war

In Hotel Hermitage in Kyiv I'm waiting for Yulia, the contact who is going to ensure that the things in the back of the Toyota get to Chernihiv. After breakfast I tidied my room and rearranged the balcony to make them fit to receive a woman – I've taken the wolf pelt off the chairs and stuffed it into the cupboard (the thing stinks to high heaven and looks like the scrawny wolfskin untrustworthy Prolix the soothsayer wears over his head and his back in *Asterix and the Soothsayer*). I've put a bottle of water and glasses ready and opened up the roadmap, and the young man at reception has promised to make coffee as soon as my visitor arrives. Adjusting to the Kyiv dress code, I'm unshaven and wearing a green jacket, a threadbare shirt, old jeans and gym shoes; it all adds up to a fairly high *M*A*S*H* content.

President Zelensky quickly emerged as an example and source of inspiration when it comes to courageous leadership, and he's also set a trend by his choice of clothing and etiquette: the simple army-green T-shirt, the unshaven chin, stout footwear, along with an absence of too much formality. It has proven infectious to visiting politicians and government leaders, who abandon protocol and suddenly present themselves at audiences in the presidential palace in Kyiv as revolutionaries, with the exception of a few stiff Americans. Zelensky's performance is strong and I believe it appeals to the average Ukrainian. Few complain about him. The Ukrainians like arguing and finding fault – a good starting point for democracy. Their president doesn't sit on a golden throne with a crown on his head, as was usual in Byzantium.

I'm a novice in a war zone, which doesn't alter the fact that a few days ago, when I was packing my things in our Hungarian village, I naturally thought about what it would be useful to bring with me. I've never been in the boy scouts, and were you to look at what I've packed you'd almost think I was trying to make up for the fact in the autumn of my life. I'm not. It's just that I wanted to be able to cope as well as possible; that was my packing logic. A garden hose with a filter, a bottle of palinka, a pocket knife with a corkscrew, a torch, a pair of compact binoculars, bottles of water, various currencies (zlotys, dollars, euros and forints), a couple of notebooks, two mobile phones, a power bar, iodine tablets, a towel, a sleeping bag, a pair of swimming trunks, a raincoat, a few shirts, and a stamped and signed document saying that the Ukrainian ministry of defence agrees to my arrival and transport. No jackets, no linen trousers, no ironed dress shirts, no dinner jackets. An error of judgement. When packing my suitcase I was thinking too much about Zelensky. The next time I travel to Ukraine I'll take a different approach. Uplifting elegance is appreciated, I've noticed, much like the way that when you're at a festival or camping in the rain you long for a hot shower and clean clothes.

In my *M*A*S*H* outfit I go down to the windowless, stuffy lobby and wait for Yulia. I was introduced to her by Illya, who lives in Poland. He knows Chernihiv well, since his grandparents live there and his father comes from there. That's probably how he knows Yulia. Chernihiv has been surrounded by the Russians for almost a month; food and ammunition are smuggled into the city at night along forest paths. I expect that Yulia, who is personally going to ensure that the waterworks in Chernihiv are repaired with the tools I've brought, is a seasoned smuggler, a lightning-quick partisan, dressed entirely in green cotton, her feet in army boots or sneakers. A kind of leopard.

I sit slouched in a deep sofa and quickly climb out when Yulia comes in. The first I glimpse of her is an anthracite-coloured ladies' shoe with a ten-centimetre stiletto heel. It's as if Holly Golightly from *Breakfast at Tiffany's* has walked in. Yulia is a little stouter than Audrey Hepburn (but who isn't?) and she lacks the tiara with fake diamonds, the string of pearls and the ivory cigarette case, but other than that, exactly the same kind of elegance has entered the lobby: all in black, with fine black stockings, a black tulle skirt with black petticoat, a black blouse, a black leather jacket, a multicoloured Hermès-style scarf and soft pink lipstick. She has a sweet face with round, blushing cheeks and big turquoise eyes bright as Alpine lakes above the treeline.

Feeling glad that I stuffed the stinking wolf into the cupboard, I lead her across my room to the balcony, to the table with the map of Great Russia, to discuss planning. The map in itself is of course an insult to Ukrainians and reason enough to lock me up. Yulia perches on a dilapidated metal chair and turns to face me. She sits bolt upright and explains who she is and what she does. She's the director of a cake factory, Kyivsky Tort. The factory is now in use by the army as a logistics centre and she helps with the provisioning of the troops. She's also a member of the local council in a sub-municipality of Kyiv, holding the foreign contacts portfolio. She's grateful I've come to support the Ukrainian army and the Ukrainian people. She says it solemnly. Her English is easy enough to follow but far from perfect. Her fingernails are perfect, though, like those of many big city women in Ukraine. In Lviv the nail studios were working flat out. It's a way – I realize as I stare at her narrow, long, soft pink, shiny nails – of showing that those bastard Russians won't get you down. Not for nothing did

Irina, killed by the Russians, the woman with the freshly manicured, cherry-red fingernails, become an icon of the cruelties of Bucha.

'My head is like a computer, it goes tack-tack-tack,' says Yulia, tapping both temples with her fingertips. 'Today we have two funerals and a lunch, and the army in Hostomel needs supplying. First the funerals. Come on, hurry up. We still have to buy flowers.'

Kyiv, Saturday morning, 28 May 2022

First funeral of the day

The first funeral is at an Orthodox church of white-painted concrete, on a wooded hill near Kyiv. I've dropped the director of the cake factory in front of the church, where the mourners are standing in groups on the square, flowers in their hands, mostly red roses. Yulia bought two beautiful big bunches of pink peonies at a stall as we were driving out of Kyiv.

I've parked the car and gone to sit round the back of the church, on lower ground, next to a huge stack of split logs. The church is warmed by firewood. It's a strange no man's land that I'm sitting in, fenced off with corrugated iron. I imagine an elderly man without a family, tolerated by the priest, who is allowed to live in this messy corner, as long as he splits wood, and is occasionally given alms. Some ragged trousers and tea towels are hanging on a washing line and there's a broken Skai desk chair on wheels and a rusting shopping trolley. Above me the hill is covered in thin acacias, the species of tree that for freemasons symbolizes the immortality of the soul. Jackdaws and crows hop around me. Behind the church is a house, no less ugly than the church. A priest and two other people move hurriedly back and forth between the house and the church's small back door. The priest in his black soutane shuffles across the open ground. The sun is shining. Dogs bark in the distance. In the square in front of the church nothing is happening yet.

It's the funeral of the son of one of Yulia's friends. The boy was billeted at Zaporizhzhia. He survived three months of war. Last week he was given five days' leave and came to Kyiv to see his girlfriend. He took her out. On Wednesday evening, a little after ten, they drove into a concrete block in the darkness on the outskirts of Kyiv. Both were twenty-eight years old; both were killed on the spot. The mayor of Mariupol has said that the concrete blocks are completely useless, that they haven't stopped the Russians anywhere, only made it harder for people to flee. They obstruct refugees and cause deaths on the roads.

I can see the upper bodies of the people in front of the church. Almost all the women have wrapped scarves over their shoulders or tied them around their heads the way pirates do, with a knot at the back. They are waiting for the family and the coffin. Death is patient and death is everywhere. Little by little, almost every family is losing someone.

A white delivery van arrives at the church, carefully pulling up at the entrance. The women crowd around it. The men keep their distance, roses held out ahead of them the way you hold a cup of coffee. The high, wailing voices of women. The doors of the van slide open. Above everything shrills the heart-piercing cry of the boy's mother, like that of a wounded animal.

Irpin, Saturday morning, 28 May 2022

Honorary Guards

We're driving out of Kyiv in the direction of Hostomel. At a crossroads with a roadblock manned by many soldiers there's a traffic jam. Immediately behind it is a burned-out filling station. Yulia instructs me to stop. This is where the transfer will take place. Everything is charred; a few steel beams are all that remain of the roof that covered the building and the pumps. We get out. There are huge black puddles of water. I take a photograph and immediately a soldier with a Kalashnikov on his belly comes and snarls at us that it's strictly forbidden. Nothing military or in the near vicinity of the military may be photographed. There is a fear of treachery and traitors. A little later a car stops next to the Toyota. We shake hands and lift the boxes out of the car's cargo space and into the back of the pickup.

Yulia suggests that she should drive for a while, so that I can take pictures. She hasn't let go of her phone for a moment all day, and even while driving she holds it in her right hand, the hand she uses to change gear. We drive along a wide road through a conifer forest and arrive in Hostomel, the suburb where the airfield is located that the Russians occupied on the first day of the war and from which they planned to take Kyiv. The nightclub and the car-wash on the edge of Hostomel were spared to some degree by the Russians, except that there are dozens of shrapnel holes in the seductive woman featured on a sign at the entrance. Practically all the houses have had their roofs blown

off, and half the front walls have gone too. In the pavements and gardens are holes a metre deep made by mortars. In every surface I see bullet holes and shrapnel impacts: in houses, fences, walls, statues.

'A friend of mine did military training at university, so she knew exactly how to pass on the coordinates of Russian trucks and troop movements. She was in Bucha. She lived near here,' says Yulia.

We drive past burned-out shopping centres, flattened filling stations, gutted blocks of flats, demolished houses and a shopping mall destroyed by a fierce fire, black as soot for hundreds of metres. In between are new flats and houses that look perfectly spick and span, with neat parks and front gardens. It's ridiculous how scenes of devastation alternate with bits of town where nothing has been damaged. The banality of those untouched houses and buildings doesn't accentuate the destruction, it obscures it, at least in places where more is whole than broken. A tall crane stands crookedly against a house, saved from falling only by the support of the building, like a crutch for a cripple.

The mobile phone has a major role to play, both for refugees wanting to know where they can go and for those who stay, both as a source of information and as a means of informing your own army about the movements and positions of the Russians, which is something Ukrainian citizens have done in huge numbers. In Bucha a Russian tank column was enticed into a trap and destroyed.

'The Russians traced the houses from which messages were being sent and raided them. My friend was fortunate in that her father, who is deputy mayor of Bucha, kept wiping everything from all the laptops and phones in the house. They took her away. Of the six people she was arrested with, two died. If they find out you've passed on Russian positions to the Ukrainian army, they torture you.'

Yulia doesn't tell me what the Russian soldiers did to her friend. As I've said, in Bucha the Russians converted a school camp into a prison and torture centre. If there was so much as a suspicion that you had something to do with the Ukrainian army or Territorial Defence Force, it meant torture and eventually a bullet in the back of the head. In Hostomel there were officers of the Russian secret service at the airfield from the first day of the war. Hundreds of men from Bucha, Irpin and Hostomel were murdered, the women humiliated to the depths of their souls. On 18 April 2022 Vladimir Putin rewarded the 64th Motorized Brigade with the title 'Guard', an honorary distinction reserved for the most capable and courageous troop detachments. A few days earlier the same honour had fallen to the 155th Marine Brigade. Those are the two brigades that operated in Bucha and the surrounding area.

In a layby next to a conifer forest are piles of rusted car wrecks. The streets were cleared after the Russians withdrew northwards. Dozens of bullet-riddled and burned-out cars were collected here and piled up. Right in front of the wrecks is a woman with long blonde hair, wearing a light-blue dress of thin material, her bare shoulders turned towards a photographer. She gracefully puts her hands to her hair. I saw the red-brown pile of cars from a distance, the woman and the photographer only later, but although we're driving at a decent speed, I manage to catch a photo of their entrancing presence as we pass.

The three suburbs of Kyiv where the Russian soldiers wrought such havoc are close enough together to spill over into each other. There are still old farmsteads from the time when they were villages, but the scene is dominated by blocks of flats about ten storeys high in sprightly colours. These were the suburbs where everyone wanted to live, with new flats and a lot of green space. According to the internet, Hostomel had 15,000 residents before the invasion,

Irpin 16,000 and Bucha 30,000. On 27 February the fight for them began. Irpin was liberated by Ukrainian troops on 28 March, Bucha on 1 April. Exactly how many civilians were killed there in the month of their occupation is not clear, because fresh graves and mass graves are being found almost every day.

We're on our way to the funeral. The bridge across which people could flee Irpin in the direction of Kyiv has been blown, the famous bridge that featured in photos seen all over the world, with Ukrainian soldiers helping children, women, elderly people and the occasional cat or dog to get across the rubble to the other side. Yulia and I have passed the bridge and are now driving over loose sand, in deep tracks, through a conifer forest, the trunks of the spruce trees broken by tank fire. The tops have been shot out of them and left to hang there like snapped flowers, the heart of the wood a very light yellow. We pass a large ditch that seems to be a mass grave. I see no bodies, only big bags the size of garbage sacks.

Last night Anastasia told me they got to Bucha and Irpin immediately after the Russians left. Bodies were lying in the streets. Anastasia is sensitive, as you can tell just by looking at her. I asked how they did that, walking along streets scattered with corpses.

'I didn't know myself that I was capable of it. I saw dead people without legs or without a head. My brain played a game with me. When I saw the bodies I'd say to myself: Is that really a dead person, that puddle of red there? Where is the head then? I'd go over to it. The head was gone, leaving only a torso. Look, the head's off, I'd say to myself, as if commentating on a football match, and that was it. You have to detach yourself from reality, otherwise you go mad.'

The bags look like human torsos. We drive past at a fair lick. I get the impression we haven't taken quite the right road. This can't possibly be the normal approach to the graveyard. It's sinister and I detect a slight panic in Yulia. She refuses to be deterred; we're on the late side. We rely on what the phone in Yulia's hand tells us. Which is to tear through the forest, along deep, sandy tracks through what was until recently a warzone. A few weeks ago there was still fighting here. In the sand are the tracks of big vehicles, and torn fragments of metal and the remains of armour plating lie everywhere. Rubbish surrounds us, as if we're at a windswept garbage dump.

The Toyota has tyres with a deep tread, for getting across the greasy, slippery clay of the Hungarian hills. The terrain here isn't a problem, but I am a bit concerned there might be mines or unexploded shells. The car weighs almost two thousand kilos, but Yulia pounds ahead. It's important, I can tell from her tense face. It's the funeral of the commander's mother.

Irpin, Saturday morning, 28 May 2022

Second funeral of the day

Suddenly there's a clearing in the woods, then a fence of green mesh with an open gate, the back entrance to the graveyard. On both sides of the dirt track are mountains of garbage, flowers, dead conifers, wreaths turning brown. It's an extensive graveyard on light sandy soil, like in a Western, as far as the eye can see. Yulia races across it at full speed. We leave a dust cloud of dry whitish-yellow sand behind us. To our left are the fresh graves, to our right, down the slope, the older ones. On that side wild juniper, pine and spruce grow among the thousands of graves. The graveyard is divided into rectangular blocks, like an American city, with sandy paths in between. Yulia has a pin on her phone, thank god, showing where last respects are to be paid to the commander's mother. She doesn't slow down. We turn right.

Below us is a white stone wall with a gatehouse. The main entrance. Yulia parks, gestures with her chin towards a group of people ahead of us and jumps out of the car with the flowers. A hundred metres away people stand in a circle around a priest dressed in black, some twenty or so, half of them in green camouflage uniforms. Yulia strides towards them across the soft sand in her stilettos. I keep my distance and walk between the graves, uphill.

Burials have taken place here ever since the Soviet Union collapsed in the early 1990s. There are erect gravestones in shiny black granite, with portraits of the deceased etched into them, based on a photo, which looks beautiful.

It brings the graveyard to life. There are a great many soldiers among them, a lot of broad-shouldered, stern-looking men. They have serious faces. Not one of the men portrayed is smiling; having your photo taken is a serious business. Almost all are in uniform. Almost everyone here seems to have died on active service.

I arrive at the central path, the one we drove in on. To my left, in the distance, I can see a tall, light-coloured block of flats sticking up above the pines. That could be the edge of Kyiv. It's so close. The Russian troops really were at the door. On the other side of the path is a new section of the graveyard. There are fresh graves, hundreds of them, in the yellow sand. A lot of rectangular pits too, ready for coffins to be lowered in. Ominous, those waiting graves. The sand is like sand on the beach and in the dunes of Bloemendaal aan Zee, light in colour at the top, darker yellow further down, from the moisture. Not a blade of grass grows there. It's an area about two hundred metres wide and eight hundred long. I look at the dates. A striking number of people died in Irpin between about 5 March and 24 March 2022. All but a few met violent deaths.

The graves have either thin black metal crosses or simple wooden crosses, reaching to eye height. Against them lean oval wreaths of plastic flowers in many colours: light blue and lemon yellow, blood red, candy pink, pale purple, orange, white, all framed in dark green. The wreaths are huge and stand almost vertical. From a distance the place looks like a Roman military camp. Some of the crosses have Ukrainian flags fixed to them with cable ties, flying above everything. People are grubbing about between the graves, relatives, family members, neatening the ribbons and staring at the elongated heaps of sand.

Separated from the metal and wooden crosses by a stretch of no man's land a hundred metres long, where scrawny grass grows, is a military cemetery. This is the place for soldiers of exceptional merit. Most of the graves feature a photo, but no headstone as yet. Each has an aluminium flagpole with a blue-and-yellow flag. The flags flap in the wind. So many dead. I walk back to the funeral of the commander's mother. It's just finished. I silently nod at the mourners as they walk to their cars. Yulia gets into the Toyota. She points the way.

As we're driving off, Yulia says, 'I do this for my heart.'

We go to Hostomel. There we take a few boxes of food to the army camp. The men are young and grateful, especially for the boxes of Kyiv cakes. Here too, much has been blown to pieces. It looks like the work of a bunch of vandals who have somehow got their hands on heavy machinery. Tanks have shot at blocks of flats, sometimes only once, leaving almost the entire building intact with just one hole and everything around it blackened, like a wound. Entire streets have been destroyed and burned to the ground; some of the houses are nothing but a heap of bricks. The garages and sheds made of corrugated iron behind the flats are often still standing, dented and burned to a rusty red-brown.

Kyiv, Saturday afternoon, 28 May 2022

Tango at Taras Shevchenko

Kyiv feels like a fashionable seaside resort off season. I keep expecting to see the sea beyond the hill, to be able to go down steps behind a shabby casino, pull off my shoes and feel fine sand between my toes, my hair stiff with salt. But on the other side of the hill is the Dnipro. Apart from sandbags in front of some buildings and the rather empty streets, there's actually not much here to indicate war. The sky is bright blue and there's a gentle breeze. I'm close to the botanical gardens, walking from Hotel Hermitage towards the Maidan, revolution square, and I let chance take me. I've had a crazy day: two funerals, a visit to Bucha and Irpin, and a delivery of food to soldiers in Hostomel. It's pleasant to wander the streets, just putting one foot in front of the other and letting my impressions sink in. Around me the houses and trees are undamaged.

Lviv and Kyiv are different. Lviv is more intimate, winding, European, Habsburg, Catholic, Jewish, with palaces built for Polish aristocrats and a lot of green space, but the most important difference is that in Kyiv the war is closer, more tangible. Kyiv is a real big city, its wide streets made for parades. Outside the old centre it resembles the Soviet Union. I actually rather like that barbaric Soviet architecture – not to live in but to look at: the insanity, the excessive dimensions intended to make people feel small, the functionalism like that of a barracks. Neal calls Kyiv a city with PTSD. On the outskirts and in the centre, close to the government buildings, it's now a fortress. The city has been under siege, with the enemy in its suburbs; it was bombed and is still hit by rockets from time to time. Were Putin to deploy a tactical nuclear weapon, Kyiv would be a far from illogical target.

I turn left off Saksahanskoho Street onto Volodymyrska Street and walk uphill. The tall houses in pastel shades are reminiscent of Beauville or Cannes a century ago: broad mansions, late nineteenth century, robust, angular, oriental, Byzantine neoclassical. The walls are pale yellow, old pink, dull orange, oxblood red, azure blue, and the decoration on the frontages, the window frames and the ornamental pillars off-white.

The older houses are three storeys high plus an attic, the windows on the third floor arched. They have a row of arches or a toothed line of masonry under the eaves, like battlements, painted white, and in the middle, high above the front door, there's a small tower or bulge. Practically every apartment has had its cast iron balcony incorporated into the interior and extended to form a sun lounge, sometimes in a way that isn't entirely ugly. There are taller buildings too, newer: Stalinist baroque. Lining the street is a broad pavement with young oak trees.

Halfway along Volodymyrska Street I pass a café with an open extension onto the pavement. It looks the way that's in fashion these days: a simple plank floor, small wooden tables, a few plants in large pots. The main difference between this and a covered café terrace in Amsterdam or Budapest is that there are no men. The lads are at the front. Fashion-conscious girls with long hair and shoulder bags are drinking Aperol Spritz from large wine glasses and looking at their phones. It makes an insipid impression, as if they're all just sitting out their time. Without men life is less jolly, there's no laughter; this is a silent terrace where nothing is said. Even gossip, the backbone of society, is suffering from the absence of men.

Fifty metres further on, two muscular lads and two girls with collagen lips walk towards me. The men are on leave and they're stiff with testosterone. They walk broadly and silently in the midday sun, radiating postcoital contentment. Their Saturday afternoon walk, hand in hand with the girls, looks like a small, polite intermezzo before they crawl under the sheets again.

On the corner of Lva Tolstoho Street I happen upon Taras Shevchenko Park. It's busy. Elderly men are playing chess under the tall trees, some with a chess clock next to the board that they hit in one fluid movement as soon as a piece has been moved. At least twenty games are taking place at small tables.

It strikes me as an excellent idea to sit in this park reading Svetlana Alexievich in the late afternoon sun, but every bench is taken. Taras Shevchenko Park is bursting at the seams. Not with drunken debauchery like the Vondelpark in Amsterdam on a beautiful summer's day. It's far more subdued than that. Drink can be sold in shops and cafés only until late afternoon. A sobriety reigns that probably stems in part from Soviet times. It's rather similar in Hungary, this socialist parochialism; what other people think of you matters, so people put on nice clothes just to go shopping in Tesco's, the way they used to for church. The war has laid an extra layer of discipline over that here in Ukraine, which I don't find unpleasant at all. People take account of each other.

There's the sound of a guitar being played in the park. A man is sitting on a bench singing softly, not for money but simply to play, to make music, to sing the praises of life. A few people watch from a short distance away.

In the Yugoslav wars Sarajevo was besieged, shelled and bombed for close to four years. My niece Ravenna has investigated how people in the city remained human. She discovered the *Sarajevo Survival Guide*, published during the siege. It's a parody, written in wartime, of all those trendy travel guides. The authors Aleksandra Wagner and Maja Razović write:

> War is not just images of soldiers shooting or the repulsive face of Doctor Karadzic or graveyards that seem to cover entire hillsides, or people and children with just one leg, war is also the life that, although more disturbing, explodes in the nights, despite the curfew, in gathering places of young men and women who look like they just stepped out of a gallery in Soho, war is also bands performing rock songs in Serbo-Croatian or 'Bandiera rossa' and 'Bella, ciao' in Italian. This is the Sarajevo spirit where Muslims, Catholics, Orthodox Christians and Jews have been living together for centuries, exchanging culture and maintaining their own traditions, a people with no intention of dividing under the deadly pressure of ethnic cleansing, that truly want to live together, there, in the city of minarets, Eastern Orthodox churches, cathedrals and synagogues, in the heart of Europe, at the intersection of East and West.

During those almost four years under siege, people in Sarajevo kept the arts alive in almost impossible circumstances. It was the apparently superfluous things that turned out to be essential. With a limited number of swimming costumes and models, a Miss Sarajevo contest was held. The swimsuits were too big, because everyone had lost weight. For the rare theatrical performances in bomb shelters, the women of Sarajevo did themselves up as if they were going to a debutantes' ball in Vienna, with the most beautiful dresses and bright-red lipstick. The frocks, the mascara, the lipstick, the performances of the bands and the poetry readings were above all a middle finger jabbed at the murderous Serbs in the mountains above them.

In *Comrade Baron* I investigated something that resembled what Ravenna studied in Sarajevo: How does a group of people survive when their very existence is forbidden? What do you pass on? How do you keep your spirits up? How do you remain human? Under communism the Hungarian and Transylvanian aristocrats had to disappear from the face of the earth. In the years after the war they were dragged out of their houses and palaces and thrown into cellars, condemned to do hard physical work, watched by the Securitate – which was instructed and moulded by the KGB – and regularly interrogated. In their windowless cellars they organized poetry readings, piano concerts and bridge evenings; they put on their last remaining smart jackets or dresses, applied make-up and greeted each other by kissing hands, using the titles, diction and idioms of a lost world. That is the essence: preserving a world and a beauty that's no longer permitted to exist and thereby demonstrating mutual ties and indomitability.

Putin has said that Ukraine is not a country and that the Ukrainians are not a people. In short, they must either become Russian or disappear into a mass grave. The ten-metre-tall statue of Taras Shevchenko, writer and father of the

Ukrainian nation, which stands in the middle of the park named after him, is covered in white sandbags. Panels have been mounted on all four sides, like those used for concrete formwork. A woman in blue overalls is patiently weeding the brightly coloured flowerbeds around the wrapped statue. A small loudspeaker stands on the steps leading up to it, playing Latin American music. On the square in front of Taras Shevchenko, two couples are dancing the tango, composed and proud, chins up.

Part III
The sniper

Kyiv, Sunday morning, 29 May 2022

Can you do that from further away?

Neal arrives on Sunday at six in the morning on the night train from Zaporizhzhia. The vast Pasazhyrskyi Station in Kyiv is visible from the front door of Hotel Hermitage, standing majestically a kilometre down the hill. In the stuffy hotel lobby we greet each other with an embrace, like old friends. He looks fit, although he's come straight from the trenches and spent the night in the train. He's in battle dress, including boots, and carrying a huge rucksack.

While he's showering and getting his strength back, I make space in the pickup. The plan for today is to drive Yulia to Chernihiv, 150 kilometres north of Kyiv. The Black Cat Cafeteria, my regular coffee bar for the last two days, isn't open yet, so we go to Pasazhyrskyi Station for coffee and sandwiches. At the entrance to the building are crush barriers and police with Kalashnikovs checking identity papers. I still vividly recall how the station looked in the first few days of the war, the chaos, panic and fear on the teeming platforms. The station concourse is like a cathedral, immense. Peace has returned. The platforms are fairly empty; trains come and go. The screen showing departure times indicates that trains are running in all directions. The many place names that light up on the screen meant nothing to me four months ago. Now they sound almost ominous: Kharkiv, Mykolaiv, Zaporizhzhia… Places that are in the news because they're under bombardment. It's 29 May. A week later, the Darnytsia train repair workshop not far from Hotel Hermitage will be hit by four rockets.

Neal and I stand in the sun in front of the station with our breakfast on the hood of the Toyota: coffee, and rolls with a white custard filling. It's pleasantly busy around us. He talks about his life over the past few months. He crossed the border only a few hours after I did – the difference between a benign aid worker in a high-viz vest and a candidate for the Foreign Legion, dressed in black. As on the Polish side of the border at Medyka, there are tents with aid workers on the Ukrainian side, only significantly fewer: one put up by the United Nations for returning Ukrainians and one by the Foreign Legion for aspiring fighters. A simple white tent. There he was given a number to call. The man who answered spoke English and explained where to catch the bus to Lviv. On arrival in Lviv he was to ring again.

An older man came to fetch him, the coordinator, fifty-five years old. He was the person who determined where 'talent' should be sent.

Neal is a quiet guy, serious, measured, not very big but well toned for his forty-eight years. He's fought in Iraq and served in Afghanistan. He was a marksman with the US Marines, and a shooting instructor and martial arts trainer. As a young marine he was stationed at Okinawa; he speaks fluent Japanese and his first two wives were Japanese. The names of his three children are tattooed on his chest in Japanese characters: two daughters and a son. He has a profound Christian faith and comes from a military tradition virtually unknown in the Netherlands, having joined the marines at an early age to fight against evil. His father was a Vietnam veteran, his mother a nurse.

He's received no payment for his work in Ukraine. The American government hasn't supported him and he hasn't asked for money from the Ukrainian government. The Ukrainian embassy in the United States let it be known in advance that they would feed him and give him a place to sleep, but that was all. Residents of his town in Alabama spontaneously decided to support him by praying for him and by collecting money. That brought in 1,800 dollars for the trip out and back, 1,000 dollars for a weapon to be bought on location, and 2,000 dollars for any other equipment he'd need. It has been estimated that around 6,000 foreigners have joined the Foreign Legion of the Ukraine Territorial Defence Forces, half of them fighters and the other half humanitarian workers and medics.

'You can get 300 dollars a month as a fighter in the Legion,' Neal says. 'And 1,000 dollars if you're at the front. Some paratroopers get 3,000 dollars. It's a matter of luck

and it depends on the point at which you arrive. I don't believe anyone has come here for the money. In Afghanistan you got 6,000 dollars a month, even if you were only doing guard duty. I haven't met many foreigners. I was with Ukrainians all the time really. I did meet an Estonian, a good guy, and a Brit, an annoying guy. They want you to have been on active service for three years, to have experience. The men who've got none at all say they were with ADC or Blackwater and insist that no one gets paperwork from them. The Brit was one of those. All talk.'

Several days after Neal arrived, he was taken by the coordinator to an improvised shooting range south of Lviv. From a distance of three hundred metres, he had to shoot at a five-inch target with a red diamond in the middle. He shows me with a curled thumb and index finger how big it was, about the size of a beermat. All five shots were inside a three-inch circle.

'Can you do that from further away? they asked. I said yes, from six hundred metres. Three out of five bullets will be on target, I told them. I did it. From six hundred metres, all five were inside a circle four inches wide. When the coordinator saw that, he took me with him to the first floor of a barracks. We sat down at a table for lunch. Borsht, tomatoes, ham, salami – I hadn't eaten so well for days. A big lad, a marine, an officer, came to sit with us. After a while he asked whether I'd ever killed anyone. I said yes. Then he asked whether I was prepared to go on a mission. I said yes. I was lucky, I was in the right place at the right time. I would be given the weapon I'd used on the shooting range, a Tikka .308.'

The Tikka is Finnish, and it's used by the Finnish army and the British SAS. In late February the Netherlands delivered a hundred precision rifles to Ukraine, the AX338, which will shoot accurately from up to a kilometre. An AX338 weighs eight kilos, a disadvantage if you're going far into enemy territory. The Tikka .308 weighs half that. Neal explains to me that in Ukraine he's killed most Russians at six hundred metres.

'I was given three portrait photos of a high-ranking officer, a Russian colonel or general. He was the target. In Lviv I went looking for a hunting store and bought a few boxes of ammunition. Boat Tail. That's what you normally use to shoot coyotes and wolves from a distance. The bullet explodes on impact, making a large wound so that you can stop big game with a single shot.'

Kozelets, Sunday morning, 29 May 2022

Kyiv cake

At the impressive oxblood-red gate to the cake factory, the same colour as on the packaging of the Kyiv cakes, Yulia and a blonde friend are waiting for us. The wide gate opens. Under Yulia's leadership, the factory has been transformed into a distribution centre for the army. A Kyiv cake or *Kyivsky tort* is the perfect gift to take with you when you return from a visit to the capital city. In peacetime the platforms of Pasazhyrskyi Station were packed with people on Sunday evenings, many of them holding cake boxes from here. It's made with meringue and it's fairly dry, so it keeps for weeks.

We load supplies for Chernihiv. The city was surrounded by the Russians for almost a month, and the only way to supply it with food and ammunition was by using smuggling routes. Two thousand Ukrainians held off 30,000 Russians with the help of the 1st Ukrainian Tank Brigade. The American general staff couldn't understand how a single Ukrainian brigade could stop a Russian tank division. The city has been declared Hero City of Ukraine.

I back the Toyota up to the cake factory's loading platform. Neal jumps into the open bed of the pickup, stands with his legs apart and takes the jerrycans of diesel from me, then the plastic storage cases and boxes with angle grinders, chainsaws and pumps. Then, with the help of the factory's security guard, we load the back of the pickup further with cans of food, boxes of Kyiv cakes and sacks of flour,

150 kilos in total, for the bakers of Chernihiv. Everywhere there are teams of women baking bread. I discover that just as Anastasia and Sasha, along with their friends, provided food to the population and soldiers in the suburbs, almost everyone is labouring away to support the army and civilian victims. Neal says that in the trenches the women from the surrounding villages bring food three times a day. The best cooks are in charge, so that the lads get the best food.

'Borsht, vareniki, pierogi – we got a lot of those dishes from the villagers. Life is hard in the trenches. You sleep in damp underground shelters and it's cold and muddy, but we ate like kings. The older women in the villages adopt the soldiers and treat them like their own sons.'

After we've finished packing and have raised the tailgate, Neal, the immensely tall security guard and I each get a round box containing a Kyiv cake pressed into our hands. We have to pose for a photo with the cake boxes turned to the camera. The Kyiv cake in festive packaging is to Yulia what blown-out Russian T-72 tanks are to foreign visitors: essential in any photo.

It's a strange side-effect of the war, and one in which I participate, this taking of photos as self-promotion. None of the foreigners I know who have travelled to Ukraine during the war can resist taking a selfie next to a pile of sandbags, a bombed building or a red-brown rusted tank, and then sharing it on social media, as proof they were here, in the war, part of history. Borrowed heroism. Life in Kyiv and Lviv isn't particularly dangerous at the moment, if you disregard the concrete blocks that are such a menace to traffic. It's more original to be photographed with a Kyiv cake than with a T-72.

I thought that Yulia's friend was coming with us just to be dropped off in Chernihiv or somewhere, but as we drive out of Kyiv I realize she's along purely for company. The pair haven't seen each other for a long time and they chatter non-stop in the back seat. Their Ukrainian sounds melodious, its chirruping turning the Toyota into a travelling aviary. On its eastern side, Kyiv is an extensive fortress. For several kilometres there are improvised strongholds and bunkers, beside and sometimes right on the road, built of whatever was to hand: slabs of concrete, sandbags, mounds of earth, truck tyres, tree trunks, welded u-bars. At the roadside and on the slopes beyond, trenches and foxholes have been dug.

We pass a huge market. Yulia sometimes breaks off her conversation with her friend to explain something to us, for example that the market we're passing is the largest in Kyiv.

The three-lane highway out of Kyiv looks entirely normal, as if we were driving along the A28 near Nunspeet on a quiet day. From time to time I have to slalom around concrete blocks or 'hedgehog' antitank obstacles. Meanwhile Neal tells me in his drawling voice about what he's experienced over the past few weeks, while the birdlike chatter continues on the back seat.

At this point fifty-eight colonels and eleven generals have been killed in the war in Ukraine, as well as hundreds of lower-ranking officers. The Russians are using unencrypted radios and telephones, so the Ukrainian army can locate them. Neal has kept me informed over the past couple of months via WhatsApp on how the Ukrainian army is training, and he's given me an idea of the missions he has carried out. The sniping he does must be very demanding psychologically. Killing people seems to me hard enough in itself, but to do it so calculatedly, calmly and quietly...

In a battle, when the enemy opens fire on you and your life is in danger, I can imagine that you naturally shoot back and have little compunction about it. But this is different. You see the face of your target through a telescope, and with the sights trained on someone's head you have to decide to pull the trigger. Neal has told me before that he can remember the precise expression on the face of every man he's killed as a sniper. After such bizarre intimacy, how do you sleep at night?

'Jaap, I was in Bucha and Irpin just after they were liberated. I spoke to civilian victims there. What the Russians have done is bestial. And that was no exception. It's the same almost everywhere. The Russians are systematically murdering and raping, in villages and towns. There's no honour at all in that army. The Russian armed forces are made up of rapists, murderers and looters. I don't see them as people. They're pigs who've made themselves ready for slaughter. They need to be taken out. I'm trained to do that.'

We take the E93 northwards. The Russians got to just short of Kyiv, to Brovary, one of the suburbs praised by Yeva and Alexey as a dream place to live. The main Russian force that was supposed to move into Kyiv from that side was stopped for a long time further north, at Chernihiv. After Brovary we drive for long stretches through woodland, then across open country again. The land is flat. Some villages have been virtually wiped off the map. Many of the houses are roofless. I drive the way I've learned from the sol-

diers who accompanied me to Kyiv a few days ago, at 140 kilometres an hour, breakneck speed. Ukraine is under martial law. People think the green pickup belongs to the army. Even police cars immediately pull over. To our left and right, rusted Russian tanks stand at the roadside, most of them missing their turrets, which were blown off when the tank's own shells exploded and lie rusting fifty metres away.

Kozelets, Sunday morning, 29 May 2022

A christening

We drive via Kozelets so that we can visit the church. Yulia has tried to dissuade me because we'll then have too little time in Chernihiv, but I've promised Tatjana Razumovsky, the friend I mentioned earlier, to check whether the Orthodox church her family built is still there or whether it's been bombed to pieces by the Russians. Kozelets is halfway between Kyiv and Chernihiv. It's also the town where Yeva's father lives with his gravely ill wife. Yeva's father helped to supply the besieged Chernihiv.

As soon as we leave the main road, taking an old wooden bridge over the Oster River to enter Kozelets, the light-green domes of the church loom up. The church is magnificent. Kozelets has a village atmosphere, with low farmhouses. During the Hetmanate, the Cossack state, it was a *sotnia* town, a settlement of hundreds of Cossacks. In the mid-seventeenth century the Tatars partially destroyed it. The little town has the good fortune to be some distance from the main road; in the past few months the Russians have not got around to repeating the work of the Tatars. We park in the main village street. Neal stays with the vehicle, since all our things are in the open bed of the truck, and smokes a cigarette. With Yulia and her blonde friend I walk through the park towards the church. Cheerful elderly village women, returning from a visit to the church, so small they barely come up to my waist, greet us enthusiastically.

As I've already mentioned, I visited this church a quarter of a century ago. At that time I was planning to write a filmscript about the battle between real and supposed descendants of the Romanovs over who had the right to call themselves a pretender to the throne – a kind of reversal of *Ninotchka* by Ernst Lubitsch. Aristocrats, unfamiliar with local customs, had returned to a Russia ruled by former communists, KGB agents and oligarchs. Tatjana was one of the people I interviewed for that project. Ever since Empress Elizabeth of Russia, the Razumovskys have been princes, and I was hoping she could tell me something about the world of the Russian aristocrats who lived as a diaspora for generations.

The white Cossack-baroque church has three towers, the middle one taller than the others, and there's a high, separate, square belltower. The roofs of the towers are domes topped with gold crosses that shine in the sun. There are three porches, one at the main entrance and one on each side of the back of the church, with pointed, tapering roofs like Cossack tents. Under the main entrance, steps lead down to a spacious cellar, where a side room houses the sarcophagus of Tatjana's distant ancestor, the mother of Cyril and Alexey Razumovsky. She stayed all her life in Kozelets, whereas most of the family left for the court in Saint Petersburg or Moscow.

'My father is very glad that you're going to look and he can't wait for your report,' Tatjana wrote. 'The Razumovskys lived in Lemeshi, a village close to Kozelets. The church was built by Rastrelli, an important architect in Saint Petersburg at the time. The mother of Cyril Razumovsky, hetman of Ukraine, supervised the building work. Her grave was paid for by Alexey Razumovsky, lover of Empress Elizabeth. It was built extremely rapidly, started in 1752 and finished in 1763. During the Second World War the church was badly damaged and it wasn't restored until the 1960s. My father travelled there in 2012, to present it with an icon painted by my Aunt Olga. It must still be hanging in the church somewhere.'

I go inside. The iconostasis is overwhelming and undamaged: four tiers of saints on top of one another, with Solomonic columns between them, swathed in golden ivy right to the top, thirty-four metres above me. Angels are everywhere. The cupola stretches up thirty-eight metres, so that it can accommodate the iconostasis. It's a church worthy of a queen.

A Ukrainian saying goes that the devil likes to hide behind the cross. That certainly holds true for Putin, who from time to time, in between murders, presents himself as a pious man. I recently read in a report by French press agency AFP that the priests of the church in Kozelets find themselves performing a moral balancing act. They come under the Moscow branch of the Orthodox Church and therefore under the insane, heel-licking patriarch of the Russian Orthodox Church. Even the ever-amicable Pope Francis has called him 'Putin's altar boy'.

Mikhail Tereshchenko, archpriest of the church in Kozelets, is a patriot. In the first few weeks of the war he made the cellar of the church, including the side room housing the tomb of the mother of Alexey and Cyril Razumovsky, available as a shelter for the villagers, who lived and slept there.

'In many of the surrounding villages, the houses were destroyed and people died,' Tereshchenko says. 'Their pain is our pain. Ukrainians, Belarussians or Russians – we're all Slavs. And the patriarch gives Putin his blessing to kill all these people. It's impossible to comprehend.'

I look up at the stunningly beautiful iconostasis, designed by Rastrelli in rococo style and carved out of limewood 270 years ago by local craftsmen. There are several people in the church; the women have covered their hair

with colourful scarves. A priest in a neatly ironed black soutane and with a spotless black *skufia* on his head comes out of the corridors to the left, a big gold chain with a cross around his neck. He has a beautiful trimmed red beard, like Vincent van Gogh. He radiates virtue in everything he does, noble in his embrace of simplicity. The priest walks to the middle of the church with a baby in his hands, until he's standing under the giant chandelier that hangs above the moveable altar with its statue of Christ, and then disappears behind the altar. Two photographers hurry past me to take pictures. Out of everyone's view, the baby is baptized.

I film the dome high above me and move my phone slowly down past the four tiers of the iconostasis to the altar, and just as I reach it a splendid woman appears, dressed in a white cotton dress that reaches to just above the knee. She's holding the baby in her arms, its head against her shoulder. The child, hair still wet from the baptism, starts to cry. The acoustics of the church are divine.

Mother, baby, husband, sister and brother-in-law pose side by side, and the photographers shoot again and again. The beauty of it is immense: the baby wrapped in white fabric, the rather ungainly, awkward men and the beautiful sisters against the overpowering background of the iconostasis, while two hundred kilometres away a war rages and a kilometre from the church the land has been churned by mortar fire. Diagonally across from me I see Yulia crying. The blonde friend wraps her in her arms. A little later, in sacred silence, they light a candle together at the altar.

We're approaching the warzone

At Kozelets we turn left along a small connecting road. The bridges near Chernihiv are being repaired, so the E93 is closed. Not far from Chernihiv is the village of Yahidne, where the Russians drove three hundred villagers into a cellar and held them there for months while the troops got plastered and looted the houses. Eighteen people died in that cellar. The villagers had to sleep among the dead, wives next to their dead husbands. On a couple of occasions, they were allowed to take the bodies away to a mass grave.

We drive along a dike across a huge expanse of marshland that surrounds the Desna River. The intact bridges are heavily guarded. There's water everywhere. Little Oster is a garrison town, where half the people on the street are in army uniform. We take the narrow P-69 northwards to Chernihiv. These are roads I recognize from the triumphant photos and videos on Telegram. Here and there we pass destroyed Russian tanks. In the middle of a large ploughed field of black earth is a burned-out Grad rocket launcher.

On the slip road the rusted turret of a tank sits like a giant turtle. *The Economist* has called the Russian T-72 the Toyota Corolla among tanks, built in insane quantities but without the necessary improvements to the gun turret and the essential separation of crew and ammunition storage, like a boxer who can give a great right hook but who has a glass chin. The thin hatch and the storage of shells under it, close to the barrel, makes the T-72 extremely vulnerable. Neal is driving, so that I can take notes.

'People told me he was an extremely aggressive commander of a tank unit that shot at civilians and non-military targets. We were given three portrait photos, one of them computer generated to show how he would look with a beard, and the coordinates of where he might be. I was accompanied by Yuri, a Ukrainian marine. He was my spotter. After I'd bought ammunition in the hunting shop and been introduced to Yuri, we left Lviv in a dilapidated Mi-8, a Russian helicopter, and flew to the south-east. That trip was the scariest of the whole war. I didn't know at that point where I was being taken. Nor had I asked. I heard only much later that it was Mariupol. Waiting for us was an old Nissan pickup, driven by two men in civvies. The Nissan took us to the front line, with Yuri and me sitting in the back. Along dirt tracks, through woods and across fields. We were to be in the area for a maximum of forty-eight hours; if we didn't find our target within that time we must get out no matter what. I think they sent in several teams on the same day as ours to kill other generals and colonels.'

The road to Chernihiv is lined with trees and bushes. It's ideal guerrilla country. There are craters in the road and at one point Neal has to drive around a rocket sticking a metre out of the asphalt. Branches hang low over the roadway; you can barely get a car past them. It's as if the road is being taken over by nature. Chernihiv is less than forty kilometres from the border with Belarus. Troops came out of Belarus as well as out of Russia. Seventy-five per cent of the city has been destroyed. Six hundred civilians died during the siege and the shelling.

'Yuri had maps with him, a compass and a satellite telephone. He knew the location where we might find the target. His task was to get me into position. We were to do all we could to avoid a firefight. Yuri had a Kalashnikov and four magazines with him, a total of 120 rounds. He was also carrying a Tokarev, a semi-automatic pistol. I had the Tikka and thirty cartridges, no helmet, no bullet-proof vest; Yuri didn't have those either. The most important thing is to be able to manoeuvre quickly and silently. I also had a rucksack with me, which held a small shovel, two sandbags, a scarf, tobacco, three pairs of socks, underpants and baby wipes.'

These are things I've never thought about before: a marksman buys his own ammunition in a hunting shop in Lviv, bullets that we Dutch don't even know about because they're generally used to hunt coyotes and wolves. And he takes clean underpants, dry socks and baby wipes with him when he goes into enemy territory to shoot a high-ranking Russian officer in the head. Neal explains how important it is not to get a fungal infection in the pubic area and especially between the toes. Hygiene is a decisive factor. Fungus on the feet is to be avoided at all costs. It's right there, between the toes and around the scrotum, that wars are lost.

'After two and a half hours we got into position. It was about nine in the evening. There was a big Russian camp – imagine a very large campsite, with dozens of small tents and three big canvas tents. Walls of sandbags had been built in front of the big tents, about a metre high. Off to one side was a line of trucks and jeeps, but no tanks. Explosions were going off all around us; there was shelling for miles in every direction. We found a place about twenty-five, thirty metres into the forest with a clear line of sight, six hundred metres from the large tents. In the dark woodland we were invisible to the troops in the camp. I first dug a long shallow trench, fifteen centimetres deep and the size of my body. I put the soil into the two sandbags I'd brought with me. I laid those in front of me, to rest the rifle on. Yuri dug himself in too. Then we took turns keeping watch.'

The bridges on the P-69 between Oster and Chernihiv were blown by the Ukrainians at the start of the war. Dikes have now been made next to them with bulldozers, so that cars can cross. As a result, large areas of land and forest have flooded. It's here in the north, on the border between Belarus and Ukraine, that the local beavers are helping the Ukrainian army. Because of the war their numbers have exploded – nobody has time to keep the population in check. They're now building dams all over the place, flooding large stretches of land and making the area inaccessible to an invading force.

As we're driving, Neal explains, like a kind of tour guide, how each form of destruction came about: the snapped trees at the side of the road are the result of tank fire, the holes in the asphalt of mortar rounds, the burned-out houses of shelling. On the edge of Chernihiv we pass two roadblocks. At the second roadblock are dozens of troops, their machine guns dug in. The mood here is clearly different from Kyiv, grimmer. A soldier checks our papers, another sticks his head inside and Yulia answers his questions. A column of army trucks with trailers rolls past. The Russians are close by. We're approaching a warzone; it makes the air quiver.

Chernihiv, Sunday afternoon, 29 May 2022

Chernihiv

Illya, the violinist I've consulted almost daily over the past few weeks about transporting supplies to the front and who introduced me to Yulia, knows this area well. His father comes from Chernihiv and his grandparents and many other relatives still live here.

During the siege in March, most supplies were smuggled over the river into the city at night. Illya's uncle was involved in that. The bus in which his aunt fled Chernihiv was fired on by Russian troops, but she survived. It's partly thanks to him that I've got here. Two weeks after I leave, I'll hear a BBC reporter say that she's one of the very first Western journalists to visit the city.

Fifty kilometres to the west is Chernobyl. Illya told me that after the nuclear disaster in 1986, his father and other students were ordered to wash all the cars that arrived from Chernobyl, without any protection, just with brushes and buckets of water. Nobody in Chernihiv was told anything. Several professors from the university were taken to Chernobyl for consultation but they were told to keep their mouths shut. In good Soviet tradition, the leadership held its cards close to its chest, the people were left in ignorance, there was lying and manipulation and a pretence that everything was perfectly under control. It's exactly what Putin is doing now: the special military operation is going swimmingly.

Rumour would have it that in the sealed-off woods around Chernobyl, dogs are interbreeding with wolves. The

local storks could no longer fly after the nuclear disaster; they ran across the fields, trying in vain to take off, and then disappeared from the area. The brown bear is back. Years after the catastrophe, Svetlana Alexievich wrote the inimitable *Voices from Chernobyl*, based on hundreds of interviews with those involved. In it she describes how, in the aftermath, Valentin Alexievich Borisevich, former head of the laboratory of the Institute of Nuclear Energy at the Belarusian Academy of Sciences, advised everyone to wash their hair with household soap, to close the windows immediately, to mop the floor every three hours (wearing rubber gloves), to put all food in plastic boxes, not to dry any clothes on the balcony and to take iodine. I note it all down because of the perpetual danger of an accident at one of the nuclear power plants in Ukraine, deliberately caused by the Russians. I'm travelling with a box of Jod Plus in my shoulder bag to give me the illusory sense of safety that I'll need should Putin have a tantrum.

Chernihiv has been attacked both by Russian troops and by troops from Belarus. Neal tells me that the Ukrainians despise the Belarusian soldiers. They were always regarded as brothers in the fight against the Russian colonizers, so those now fighting with the Russians are the ultimate traitors. The Russians wear thick green coats, the Belarusians black jumpsuits. When they leap out of a truck or armoured personnel carrier together, the Ukrainians always aim first at the men in black.

About half the houses along the road have been destroyed. Even after the Russians left, Chernihiv was regularly bombed, right up to twelve days ago.

Shortly after the siege was broken by the Ukrainian army, Yulia drove to Chernihiv with a truck full of food. The city was almost impossible to reach at that time. She shows me a video on her phone. It was shortly before Orthodox Easter. You see throngs of people in cheap winter coats against the background of destroyed blocks of flats. One woman falls into Yulia's arms in tears, having been handed some bread. In a Russian bombardment of a bread queue in Chernihiv on 17 March, fifty-three civilians were killed. Before the war the city had 300,000 residents. Half of them have fled.

We stop at a filling station. By this point I'm used to the fact that things are transferred from one vehicle to another at petrol stations. A chain of volunteers helps to move people, jerrycans, food, medicines, weapons and ammunition all over the country. A little later a delivery van pulls up, carrying an alderman from Chernihiv and the coach of Desna, the local soccer club, which plays in the Ukrainian premier league. We drive through the city behind them, passing a park, its grass closely mown. The sun is shining, people are out for a walk. Everywhere I look I see beautiful churches, white and pale yellow. At a few rare places there is no sign of the war.

'This is a city of churches,' Yulia says.

We stop on the square in front of the football stadium and get out. It has an imposing entrance, with tall square pillars in Soviet style.

'Chernihiv is one of the few cities the Russian air force dared to bomb,' the football coach says. 'Within two minutes they're back in Russian airspace from here.'

He points to the centre spot. The Russians bombed a number of attractive nineteenth-century buildings, and absurdly enough there's a bomb crater in the middle of the main pitch, as if the Russian air force wanted to show how precise its bombing could be. Or maybe it was the work of a magnanimous Russian pilot who had to drop his full load and choose this as the place in the centre of Chernihiv that would do the least harm.

We continue our journey. We still need to deliver 150 kilos of flour to the women who bake bread for the city. We drive through a district that's been shelled to pieces. Along the road are more burned-out trucks and T-72 tanks. We take a little street lined with fruit trees that could be any Eastern European village street, were it not that every house is at least half demolished. There are large piles of rubble at the roadside.

It's May, everything is sprouting, the trees are blossoming, the grass in the verges is fresh green, and at most of the houses the residents are clearing rubble. Next to a house of which only the brick walls remain is a burned-out Ukrainian tank. Through a hole blasted into a wall I see an old fireplace, with a white teapot at the edge of the mantlepiece. I inch into the verge, under an apple tree. The grass here is covered in shrapnel and other bits of metal. I place my feet carefully, for the first time in my life afraid of stepping on a mine. Chernihiv is famous for the many mines the Russians left. Patron, the decorated landmine detector dog, found more than two hundred of them here.

A fat woman in tracksuit trousers walks across the garden and gestures to me to come over. She goes ahead of me to the skeleton of a house. We clamber across piles of bricks. A filthy smell of fire still hangs in the ruin.

'This was my grandfather's house,' she tells me. 'The Russians shot it to pieces.'

In a corner of the living room, three shell casings are neatly stacked up. All the rooms have debris a half-metre deep on the floor. Through a back door we walk out of the ruin. Here, on planks laid on trestles, piles of material retrieved from the rubble have been sorted: steel with steel, stone with stone, wood with wood. In a corner is a blackened heap of bricks. Next to the house are four recently constructed walls, making a long shoebox of four by eight metres in white Ytong blocks.

'Our new house,' the woman says proudly, pointing to the white structure.

The Russian and Belarussian troops are regrouping a few tens of kilometres north of the city and these residents of Chernihiv have almost finished building their new house. The big white sacks of flour have meanwhile been lifted out of the Toyota and taken to a large outbuilding in a back garden. We drive on to the waterworks, where we deliver the twenty tools I bought in Hungary. The recipients are grateful and want to get started right away. These people are unsinkable. In February, 56 per cent of the Ukrainian population believed in victory. By Independence Day in August, that will rise to 93 per cent, according to the polls. On leaving Chernihiv I see people painting the snow-white rectangles of a zebra crossing on fresh asphalt with big brushes. At the roadside are mounds of rubbish from bombed houses, raked together. A man is mowing the grass.

On the way from Chernihiv to Kyiv, Sunday afternoon, 29 May 2022

Make sure they don't take you alive

Neal speaks very calmly, without any sign of bravura, factually and with precision. He's driving and I'm in the passenger seat. He turns towards me and draws a triangle on his face with his index finger.

'You can draw a triangle between the pupils of your eyes and the tip of your nose. You aim for the centre of that triangle, for the bridge of the nose. From there the bullet goes to the centre of the brain, to the medulla oblongata, the brain stem. Hit that and the human body goes out as if you've flicked a light switch.'

For the first part of the journey out of Chernihiv we take a different route, crossing a half-collapsed bridge of which one traffic lane is still passable.

'We were very lucky. We'd been waiting for less than eight hours. At first light, a man appeared out of one of the big canvas tents, dressed in a white kimono. Under the kimono he was wearing the blue-and-white striped undershirt of the Russian military. He stepped out of the tent holding a steaming cup of coffee. He had two weeks of beard growth, neatly tended. He looked like the target. My spotter confirmed it was him. Yuri compared the photos one more time. He had a telescope with sixty times magnification, my telescope gave me only twenty-two times. Yuri said: It's him, hundred per cent certain. It went quickly. Less than three minutes after he stepped out of the tent, I shot him. I don't know who the man was, I don't know whether it was a colo-

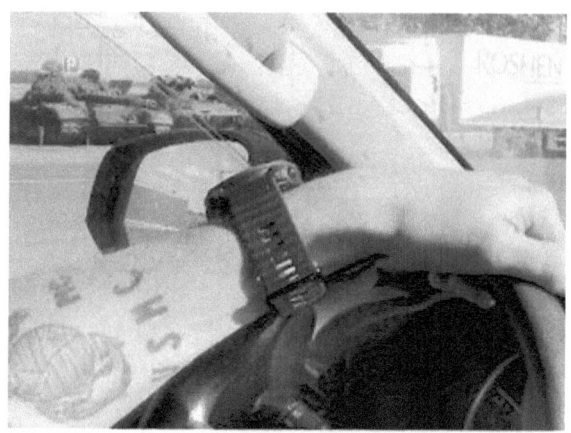

nel or a general, but I do know that his men will have been terrified when they crawled out of their tents.'

The rapeseed crop is flowering; we're driving between vast, waving fields of yellow.

'After the shot we pulled the thermal blanket over us and I reloaded the rifle. If other people appeared, I would shoot them too. We kept scanning the camp, I through the rifle sights, Yuri through the telescope. Nothing happened at all. Everyone was peacefully sleeping. Nobody had noticed anything. They felt so safe that they hadn't even posted guards. After a few minutes Yuri and I packed our things and went through the woods to the closest recovery location. There were several places arranged and each had its own code. Yuri used the satellite phone to say we needed to be picked up at the Alpha spot. We waited at the edge of the forest. The same helicopter came to fetch us. I was very relieved when we jumped into it and the decrepit thing took off.'

His shirtsleeves are rolled up. Both his arms are covered in tattoos that reveal the fact he's a US Marine. There's an eagle over a globe and above it the Marines' motto *Semper Fidelis*, 'Always loyal'. Next to that are tattoos of Band of Brothers, Leathernecks, and USMC, with similar images. All in blue-black. If he was captured it would be impossible for him to conceal his identity. Neal is a dream trophy for the Russians. I ask him what he'd have done if they'd surrounded him.

'I'd have made sure I didn't fall into their hands. That they didn't take me alive. I've got a knife. I've got hands.'

As a Christian, suicide is not an option for Neal. He would fight to the death. A considerable number of foreigners in the Ukrainian Foreign Legion have already been killed, more than a hundred. Foreign fighters are honoured on the square in front of Saint Paul's Church in Kyiv. I saw these numbers there: 1,000 Georgians (the Georgians have provided the Ukrainian Foreign Legion with the largest number of fighters), of whom 50 dead; 400 from the United States, one of whom has died in battle; 50 from the Netherlands, two of them killed; 250 from Belarus, 100 from Morocco and 100 from Britain. In total there were 6,000 names on the board on the square, men from 50 countries. Two Brits and a Moroccan have been captured by the Russians. All three have been sentenced to death.

It sounds nerve-wracking to me. You walk through enemy territory for two and a half hours, across land teeming with Russians. You know that if you're captured you'll be flayed alive. You arrive as night is falling. You kill a general or a colonel and then it gets light. There are soon probably hundreds of Russians looking for you, and if you're unlucky they'll be using helicopters and drones. I ask him what the moment after the shot was like.

'Yuri said he immediately collapsed like a pudding. He disappeared behind the sandbags. I couldn't see the man any longer, because of the recoil of the rifle, but I heard the shot and heard that it was on target. The shot always makes a noise, even with a silencer. As the bullet leaves the barrel it gives out a whooshing sound, *woof!* And you hear the impact.'

I know from big game hunting that experienced game wardens can tell whether a shot is on target, whereas I hear nothing. I ask what kind of sound it is.

'It's...' He hesitates for a moment. He thinks about it. He holds the steering wheel with both hands and stares straight ahead. '*Flap!* As if you'd hit an uncooked turkey with a baseball bat.'

Kyiv, Sunday evening, 29 May 2022

In Russia sin is as boring as virtue is among us

I'm standing in the back of the Toyota, sucking noisily at a garden hose full of diesel. At the end of each day I fill the tank, so that the next morning, or even during the night if necessary, I can drive off without delay. In Lviv I developed the correct technique: jerrycan on the roof of the driver's cab, then climb into the back of the truck and suck the diesel through the hose. When the hose is full to thirty centimetres from my lips, I bend the end double and then poke it into the tank. This technique works well, except that when I have to start on a second jerrycan, after doing a lot of kilometres like today to Chernihiv and back, I have to put the end of the hose into my mouth again. The diesel taste is hard to toothbrush away.

Every evening I calculate like a minor bookkeeper how many more kilometres I can drive. I have to make sure I've got enough fuel left to make it to the Hungarian or Polish border. At the few filling stations that still have supplies, the queues are endless, and there's no diesel on sale there anyhow. It's available only from hustlers in delivery vans at the side of the road who have placed a jerrycan next to the vehicle or on the roof, as a sign. Dirck has told me that in case of emergency I can fill the tank at his farm, between Zhytomyr and Lviv, but I don't want to deprive the Ukrainian tractors of their fuel.

I'm busy with this intelligent siphoning job when suddenly there's a man standing next to me. It's philosopher and psychologist Vlad Beliavsky, the author of the as yet unpublished *The Pyramid Mind*. I suspect he's appreciative of finding a Dutch writer standing in the bed of a Toyota pick-up with a concentrated look on his face and a garden hose in his mouth. Vlad is the stepbrother of Dirck's stepsister.

What on earth's going on? How can it be that I've met hardly any Ukrainians whose parents are still together? They're almost like the Hungarians, who leave each other for the flimsiest of reasons. Ilona, my wife, used to run a business in Budapest that employed around sixty young Hungarians. They partied loud and long, married and just as rapidly divorced again. Of the twenty weddings among her merry workers, only a few heralded lasting marriages. It must be a lingering effect of communism. The sexual morality encouraged from above under communism in Eastern Europe – which focused on conceiving children without sentimentality or romance – and the compulsory company vacations to the Crimea or Lake Balaton or wherever, held in big holiday venues owned by the company's branch of economic activity, with male and female employees crammed together for a week of sun, swimming togs, drink and dancing, were mainly intended to break what remained of that last nucleus that could still provide a counterweight to socialism: the family, with its wedding pledge. Under communism there was a taboo on property (resulting in a secret obsession with material belongings), and the possessiveness of marriage was to be done away with as quickly as possible. The complete atomization of society was the goal. It suited the omniscient state if there was a lack of trust between married couples and they were exchanged as easily as dance partners. This resulted in complex family structures, with stepmothers, stepfathers, stepbrothers, stepsisters and so on.

I'm reasonably familiar with how it goes: the involuntary doubling of the number of family members, the sensitivities of having a stepfamily, and the awkwardness of such constellations, from which the children are often the ones who suffer, even if the adults claim it's all going fantastically well. My mother left my father when I was two years old, and to keep it all in the family she married the man whose former wife my father then married. A structure you'd expect in the utilitarian, post-war Soviet Empire, not among 1960s manufacturing families in the Dutch province of Twente. By the time the new marriages were in place, I had a half-brother, three brothers, a stepsister, and four stepbrothers, of whom the last five were doubly step. Of the ten of us, I was the youngest but one and expected to keep my mouth shut, which I did.

I don't believe I'm giving away any major family secrets if I add that my mother and stepmother hated each other's guts. Family parties and seating arrangements needed to be devised with some circumspection. It was all very civilized, certainly, and the warring factions limited themselves to sly digs, with no Sicilian throwing of crockery, alas. As a writer I could at least have made something out of the spectacle of flying Wedgwood or Gien Oiseau Bleu. However that may be, as a structure our extended family would have fitted neatly into a socialist Eastern Bloc workers' paradise, with the accompanying kitchen whispers.

Travelling through Russia and Ukraine in 1926, Joseph Roth foresaw the consequences of the socialist legislation that reduced love to a hygienic, immaculate pairing between two individuals sexually enlightened by school pictures and brochures. 'Sin in Russia is as boring as virtue among us.' In an article published in the *Frankfurter Zeitung* on 1 December 1926 he wrote:

> *Marrying is easier than reporting a crime to the police. Husband and wife work and have meetings all day long in separate companies. If they happen to discover on a Sunday or at a demonstration they both attend that they don't suit each other, or if one of them would prefer someone else, they divorce. Man and wife know each other even less well than partners in a capitalist wedding-with-dowry. Divorce is more frequent than among us, because marriages are entered into more 'frivolously', less thought through beforehand.*

Writer and philosopher Vlad has divorced parents. Anastasia, the young musician, has a Russian father and a Ukrainian mother, who despite being former bedfellows could happily shoot one another, further encouraged by a murderous nationalistic mutual aversion. Sasha's mother was single. Margarita, who fled to Budapest on the sixth day of the war and for whom I'm going to take back three large bags of clothes, has a Russian mother and had a Ukrainian father who divorced as if it was the most natural thing in the world. She herself has already been married and divorced once or twice. I didn't ask Yulia of the cake factory, but it wouldn't surprise me if a marriage with an unhappy outcome is slumbering somewhere there. Ella, the barista who brought the three girls from Kharkiv together and guided them to the border, was divorced and had left her young son with her ex in Ukraine. Yeva's mother has a second husband who is now stuck in Zaporizhzhia, where he has to look after the children of his divorced son. Yeva's father has also remarried, and until recently he was stuck in Kozelets. The pleasant Polish aid worker Michael, whom I met at the Medyka border crossing, had never even met his father until the war broke out. The only person from whom I've explicitly heard that his parents are still together and undertake all kinds of things as a couple is Illya, the Ukrainian violinist in Poland, but other than that it's one endless circus. Neal is a child of divorced parents too, and he's been divorced twice himself. Perhaps wars hold a special appeal for children of divorced parents. Perhaps unconsciously they hope to be able to repair, amid the chaos, something that always seemed irreparable before.

Kyiv, Sunday evening, 29 May 2022

The Pyramid Mind

I run the back of my hand over my lips as I walk with Vlad along Symona Petlyury Street to the Georgian restaurant close to the hotel. He's full of energy. I know him only from the photo Dirck sent me, crouched in camouflage fatigues, stroking a fox-like dog. Neal joins us a little later. He orders whisky with ice and water. I make it two; it washes away the taste of diesel. The Georgians have a casual attitude to the after six drinks ban, thank god. Vlad sticks to water. He needs to rejoin his unit this evening.

Vlad has been an officer in the Ukrainian army for several months. He tells us how in Ukraine the younger generation has made clear to the older generation that they want everything to change. At the revolt on the Maidan in 2013, young people played a major role. They still do. Many of them are in the army. At the town hall in Lviv I saw memorials to dozens of dead soldiers and none were older than twenty-five. In Russia and in many Eastern Bloc countries there has never been such a revolt by the young; older people still determine what happens and the habitual Soviet corruption rumbles on.

Vlad discusses with Neal the moral aspects of being a soldier. It's a conversation between two military men, one dyed in the wool, the other just starting out, one experienced at killing, the other still a virgin in that respect. It's a mild evening and we're sitting at a small round table on the terrace in front of the Georgian restaurant. There are still two other groups of people here. The atmosphere is relaxed, almost cheerful, certainly around Neal, who is on his way home and has survived it all. On the broad road before us there isn't much traffic. The black of the night is still hours away.

'Did you want to hit him the moment you took aim?' Neal asks.

'If we could have got them without killing them, that would have been better,' Vlad answers. He explains that in the early weeks of the war he was ordered to patrol the north-western side of the city, close to Bucha and Irpin, to intercept spies and teams of saboteurs. The FSB has invested hundreds of millions over the past few years – in Ukraine the talk is of an improbable sum: five billion euro – to sow division among people in Ukraine, to incite the minorities (Russians, Romanians, Tatars, Hungarians and so on) to

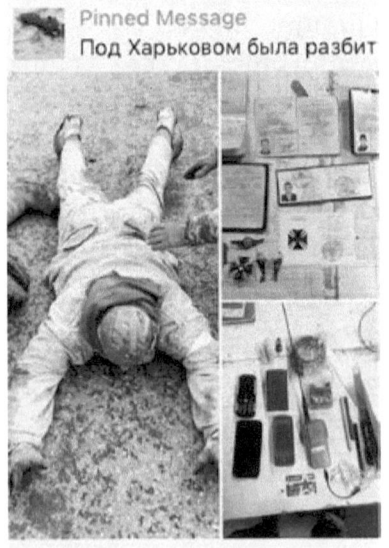

Pinned Message
Под Харьковом была разбит

В Николаевской области "освободители" оказались в плену

form a fifth column, including at government level, and to prepare people for a Russian takeover. Assassination squads were sent in to kill Zelensky and all the top people in government. The Russians were only a few kilometres from Zelensky's office.

FSB agents rented a large number of apartments in Kyiv that people were told to abandon in the days before the invasion without locking the door, so that new FSB officers, arriving with the army, could move into them during the takeover of Kyiv. The FSB had two different groupings ready that could form a new government. The necessary reinforcements and staff would be flown in to Antonov International Airport at Hostomel, north-west of Kyiv. The Russian elite troops that took the airfield on 24 February were accompanied from the very start by FSB officers. All that stood between the Russians and the government quarter were Ukrainian soldiers and militias, who fought to the death, and the River Irpin, which overflowed its banks after the dam further upstream was opened. When the Russians, contrary to their own expectations, failed to advance into Kyiv immediately from Hostomel, they still hoped to be able to force entry via the village of Moshchun further to the north.

In those first few weeks, Vlad led a patrol in the districts of Pushcha-Vodytsia and Obolonskyi, where Yeva and Alexey have their flat and Yulia her cake factory. Neal asks how they managed to identify saboteurs.

'You could tell from their behaviour, their fake papers and transparent lies. They struggled to find their way through the city, and that often gave them away. Plus the fact that they tried to avoid us. We often patrolled in the evening or at night, and we'd stop everyone then. Of course we could also pick out the Russians by the way they spoke; there are a lot of words they can't pronounce. We checked their phones, too, and if they had photos or coordinates of our positions or military targets, you knew they were up to no good.'

Vlad seems to me the ideal person to detect liars. He's a young man with a sharp intellect. Freedom, free will and responsibility are subjects he's worked on for ten years. He studied IT in Ukraine and did military training at the university alongside his studies. He went on to study philosophy and psychology at the University of Warwick and got his PhD. At the age of twenty-one he wrote his first book, called *Freedom, Responsibility, and Therapy*, published by the renowned academic house Palgrave Macmillan.

There are all kinds of Ukrainian words that the Russians can't pronounce properly, just as the German occupiers of the Netherlands in the Second World War couldn't say the place name 'Scheveningen', which was therefore used as a test. 'Palyanytsya' is one of the best-known words, a type of bread and unpronounceable for Russians, who also have difficulty with the words for 'pheasant' and 'cloudy sky'. Someone says, 'The sun isn't shining today', and you're supposed to answer, 'Yes, it's a cloudy day.'

Neal explains how people recognize each other as belonging to the Ukrainian army. Russian soldiers often pretend to be Ukrainians, so the Ukrainian troops have blue or green tape around their arms, and on their vehicles. The colour of the tape changes every week or two. Every few days a new code is given out, such as 'minus two' or 'plus one', and distributed throughout the Ukrainian army. At checkpoints or if you come upon troops you don't know, you stick up five fingers or whatever number you choose. The other person has to respond with the same number of fingers, adjusted according to the code for the day. So if the code is plus three and the man at the checkpoint sticks two fingers in the air, then you answer with five raised fingers. Neal adds that in the American army it's much easier and doesn't have to be arranged centrally. The check usually involves sporting heroes. 'First base for the New York Yankees?' or 'Who's the top scorer at the Chicago Cubs?' These are things practically everyone serving in the military knows.

Vlad tells us how he decided to sign up with the army.

'It was on 4 March. I'd given it a lot of thought. I'm not afraid of dying. My greatest hesitation concerned whether I was prepared to kill someone else. That was my big inner conflict. I've hardly ever been angry in my life, never felt that urge. I'd never expected to join the army, but when I saw the Russian army firing on the Zaporizhzhia nuclear power station, Europe's biggest nuclear plant, I became convinced this madness had to be stopped. Putin is a psychopath, just like Trump is a sociopath. I had to accept my responsibility. 4 March was the day of my decision. Our world could be destroyed at any moment. I couldn't just stay sitting on the sofa doing nothing. If the nuclear plant at Zaporizhzhia blows up, large parts of Europe will become uninhabitable. I have to do my bit to try to prevent that.'

Vlad looks at Neal, at his decorations from various units and at the 'master hunter' medal on his chest. He wants to know how Neal deals with stress and with killing people.

'I come from a religious family. I'm a member of the Methodist Church and I pray every day. On Sundays I put extra time aside for reflection. I spend that whole day in gratitude, for my family, for friends; I'm grateful for ordinary things. As far as days at the front are concerned, I have simple routines, I don't focus on the battle but live from meal to meal, and through the week I live from church to church. Hold on to your principles and pray – that's what helps me. And when I kill, I don't see my targets as people. I turn the enemy into animals. They need to be taken out. People who fight under the Russian flag are terrorists. There's no sense of honour in that army.'

Neal takes a swig of whisky. He looks troubled. He grasps the rosary around his wrist and turns the black beads between thumb and forefinger, one bead after another, as if they're the heads of the men he has killed. He's told me that on sniping missions in and around Mariupol he took out fourteen Russians, most of them in leadership positions. He sits leaning forward, eyes to the ground.

'How can I face what I've done?' Neal straightens his back and sits erect, looking at us. 'It's a religious process for me. They are very much alive. They are very human. And now they are dead. I send those people to God. When I see them in the sights of my rifle, I realize that I'm about to send them to their Creator. And I can see them very clearly. Their faces, their expressions, their eyes. I prepare them for the Last Judgment. And that judgment is God's.'

We drink in silence. Neal runs the beads through his fingers.

'I've got this, a rosary.' He shows us how he lets it slip through his hand. He nods at the ground. 'Just feeling it helps. Especially if I count at the same time. That's my way of having control. The most difficult thing in a war is the sense of helplessness. The feeling that you have some control is essential. It's horrible in the trenches and underground bunkers when the Russians are shelling you. You can only hope that nothing lands on the place where you're sheltering. It helps at that point to be focused on something. I remember being underground, with tree trunks above us and just one or two metres of earth. Everywhere around us, rockets were impacting. By focusing my eyes on a knot in the wood and seeing the beauty of it, I pulled myself through the madness. Being in command of your head is the trick.'

'As a people we've got to the stage now where we hate the Russians and want to kill them,' says Vlad. 'Yes, it's difficult to stay calm and keep thinking when rockets are coming in all around you. You lose your sense of time. Even as a soldier, I do my best to remain human under all circumstances. The invasion filled me with horror. Putin thinks the Ukrainians have no right to exist, that Ukraine isn't a country and we should disappear. But despite everything I try to be free of judgement and to stay in the here and now. To observe, use my senses. Smell, feel, see, listen.'

Neal nods his approval.

'We need mentally healthy people,' Vlad goes on. 'I try to teach my troops mindfulness, so that they live in the present moment, in contact with their senses, and learn not to be judgemental. To maintain inner peace. It works. The young guys adopt it from me. Many are no older than twenty. I try to help the lads to grow. I've given a lot of thought to how we can make the world better. I studied philosophy and psychology so I could make a contribution to that. I thought: if we have good leaders, it'll be fine. But we live next to an aggressive pit bull and on the other side is a vacillating French poodle. We want our freedom. And of course we all have one ultimate goal: to survive.'

On the way from Kyiv to Zhytomyr, Monday 30 May 2022

On fear

There are dark clouds when I leave Kyiv for Zhytomyr on the morning of Monday 30 May. Driving in the direction of safety, I already miss the excitement, the solidarity, the kindness and gratitude of the Ukrainians. In the suburbs of Kyiv, in between the steel antitank obstacles, are street sellers with onions, potatoes and strawberries, and on the pavement elderly women sell cut flowers from plastic buckets. The E40 is scattered with concrete blocks and with makeshift bunkers thrown up in the middle of the road. A few of the roadblocks, which you have to zigzag around, are manned by troops in bullet-proof vests with Kalashnikovs, but most are deserted. Neal is travelling in his camouflage uniform with his Ukrainian medals and decorations, which pleases me, because it ensures us of priority treatment.

At checkpoints he's greeted with '*Sláva Ukrayíni*' (Glory to Ukraine) to which he gives the standard response, '*Heróyam sláva*' (To the heroes, glory).

This brief dialogue is significant in its simplicity, since it evokes the desire to be one of the heroes yourself. It's of a quite different order from the Dutch and American 'Have a nice day'. We Westerners strive for comfort, consensus and ease, and wish each other exactly that. A nice day is our highest good. We're no match at all for the Russians with their violence and love of suffering, but fortunately the Ukrainians are. Here they wish each other heroic deeds. Several people, often young, have said to me over the past few days that they're prepared to die and are not afraid. The effect is contagious. It's also a matter of willpower, of deciding, consciously or unconsciously, not to be afraid.

I ask Neal whether he has felt fear. He gives a long answer, saying that it's not so much fear as an enhanced state of alertness that he enters, very much aware of the danger, all the time but especially when he's on forward operations. He describes the raised adrenaline levels. I recognize that, even though I've always been a long way from the front line. After arriving home in Budapest I'll retain my state of arousal for two or three days. It's that, I think, that makes wars addictive – the intensity of life, like being in love.

My eighty-eight-year-old mother, who at the age of eight possessed sufficient cool-headedness and sagacity to warn people hiding in a wall cupboard in the Schopman farmhouse in Twente that this was the moment to come out and disappear into the woods, while above them a squad of Wehrmacht stamped through the house in search of illicit inhabitants, told me not long ago that she never actually feels any fear. 'I can be worried, but something pretty crazy has to happen to make me afraid,' she said. She's certainly hard to unsettle. Perhaps her attitude is the consequence of an old-fashioned, privileged childhood.

Erzsébet T., a ninety-three-year-old Transylvanian countess I interviewed many times, said she was never afraid simply because her parents had never taught her to be. 'Fear is something you learn from your parents, from your nannies.' Of the interrogations and torture inflicted on her in the early 1950s by the tyrants of the AVO (the Hungarian secret police), who tied her to heating pipes and one time broke her fingers, another time her toes, she said, 'Everything you experience in your life is interesting.'

Erzsébet's attitude is admirable. To remain astonished and interested while your fingers are broken or cut off seems to me superhuman. But I like that stoical approach and something of it was passed on to me during my upbringing. When the air-raid siren went off in Kyiv or Lviv I simply couldn't imagine that a cruise missile or a piece of shrapnel would hit me. In my days in Ukraine I was worried only when Yulia was driving through the forest near Irpin, because there might be mines, and when Neal was at the wheel and overtook by snaking between cars on narrow roads.

To our right lies marshland. We're driving past Kopyliv. Bucha and Irpin are behind us. It's still raining. The sky is completely grey. Beside the road are charred filling stations. The houses are pulverized and unharmed by turns.

'I've got an army pension of 3,000 dollars a month,' Neal goes on. 'I was blown up in Iraq. We were with four other vehicles on our way to meet a convoy, on the main road between the west and the east, between Bagdad and Fallujah. There were four of us in a Humvee. I was sitting next to the driver when an IED exploded. The driver was killed on the spot, the man behind me lost an arm. I was only lightly wounded, but after that I had memory problems and the army gave me a medical pension.'

For the first few hundred kilometres in the direction of Lviv, the road is a dual carriageway. I let Neal drive here, because he drives like an American, overtaking with insufficient aggression and acceleration. On the road to Chernihiv he crept past big trucks, which is nerve-jangling on a single-lane road. The rest of the road users here drive with Eastern European machismo and contempt for death. Giving way to oncoming cars means loss of face. So I make sure I'm the one driving after the road narrows at Rivne, when the gas pedal needs to be floored as you overtake.

Neal's parents divorced when he was a year old. His father, the Vietnam veteran, was not an easy person, whereas his mother, the nurse, was motivated by her caring nature. His father is no longer alive. His mother now runs a hospital consultancy; she's an entrepreneurial woman. I can hear that he's proud of her.

'I liked adventure and used to get into fights as a child. The Marines were good for me. My mother was very worried when I came here. I told her, I'm not going to fight the Russians, I'm going to protect the women and children. That put her mind at rest. And that's what I've done, too. After I took out the target near Mariupol, they asked whether I wanted to perform further missions. I said yes.'

The 'ordinariness' of war, of the environment, still surprises me. The difference it makes – the ruins, the black stumps, the roofless houses, the charred buildings – is actually minimal amid all the green. The surroundings we're driving through now are identical to the world in which I live. Undulating fields. Linear settlements. Modest farmhouses with fruit trees, vegetable plots, stacks of firewood, and collections of cast-off items that might come in handy one day. The shopping malls on the American model, with their huge parking lots, big rectangular structures rapidly thrown up, with corrugated roofs and walls, have in many cases been gutted by fire, leaving only the vertical and horizontal steel beams. The Western companies that have settled onto the outer edges of Eastern European cities, besieging them like an occupying force with supermarkets, hardware stores and electronics chains along the slip roads, seem particularly prone to being shot to pieces. When peace eventually comes, they'll quickly be reconstructed. Generally speaking, people in the destroyed villages are worse off than people in the towns and cities; the villagers mostly own their uninsured houses, whereas a lot of city dwellers rent.

'We operated in a group of four, a good group, two Ukrainians and an American who'd worked in international security. The Ukrainians took us everywhere and covered us, making sure no Russian patrols bumped into us. It was a target-rich environment, with lots of Russian patrols, sometimes of six, twelve or sixteen men. The orders were to shoot targets of opportunity. I sometimes fired from blocks of flats. A building is one of the most dangerous places for a sniper, so I never shot from higher than the second or third floor. When you go inside you keep a 3D image of the building in your head. We would set up a table, deep inside a room, several metres from the window, and fill sandbags with rubble to rest the gun on. My spotter, the American, took account of the wind speed and the distance and helped to choose the target. In the last eighteen hours I got greedy, too greedy. I normally shot the man in charge. I'd fire and then we'd move on. Fire, move, fire, move. But towards the end I shot four times from the same place. I was pushing our luck. They didn't call in any artillery, they didn't move towards us, they didn't even try to find out where we were. They ran.'

A pleasing number of heavy army trucks are driving towards us, heading east, from Poland towards the front. Semi-trailers, both army and civilian, carrying tanks and armoured personnel carriers, often tightly wrapped in plastic, concealed from view. Usually in convoys of two or three, a car with flashing lights in front and behind.

'After the sniping missions I went from unit to unit to train soldiers, first near Kyiv, then near Zaporizhzhia,' Neal says. 'The army is largely made up of men and women who a few months earlier were still doctors, lawyers, shop assistants or computer programmers. They're like sponges, they suck up all the information, convinced as they are that it can make the difference between life and death. They live in trenches, in huts deep under the ground, often ingeniously built with retention pits for rainwater.'

The kilometres-long, thin corrugated sound barrier next to the highway has been fired at and bits of it have blown into gardens, woods and verges. Water from the road surface splashes up, reducing visibility. The rain taps on the roof, reminding me of my childhood, the cosiness of a wet

holiday, a tent in the autumn vacation on the expanse of the Beulakerwijde lake or the meadows of Herr Müller's farm in the Sauerland.

'In Iraq we weren't allowed to have mobile phones. You had to wait in line for hours to call home for ten minutes. Now I can send photos via Signal and make video calls. That's the best thing; I can see my wife smile. I can see the relief on her face when I get through to her. I can see my children, speak to my son. That's great. That makes this war far better, the contact with home. I'm very grateful to the Ukrainians. They're warriors. Ukraine has given me a lot.

'When I came back from Iraq I had a rough time. I rang my father, wanting to know how he dealt with posttraumatic stress after Vietnam. He'd had comparable experiences, fifty years earlier. All he had to offer was macho bullshit. "If you want to talk about this kind of nonsense you need to go to a doctor and not call me." So I said to him, "This is your last chance to make contact with me, to show that you're my father." He just hung up. I never spoke to him again. I didn't go to his funeral. I still quarrel with my brother about that.'

On the way to Lviv, at Brody, Monday 30 May 2022

A boy called Samuel

Galicia, the western part of Ukraine, up to a short distance beyond Brody, belonged to the Habsburg Empire for a while, which gradually developed it, laid railways – that are now being bombed from time to time – and enfolded it within Habsburg rule. The Habsburgs were happy to endorse the Greek Catholic Church, shrewdly imported into these parts by Polish landowners, which followed the rituals of the Orthodox Church yet recognized the pope in Rome as its supreme authority. All minorities were allowed to keep their language and faith. The basic approach of the Habsburg Empire was always to keep the peoples within its extensive territories equally happy (or unhappy) and where necessary to play them off against one another.

'Shortly before I was due to leave Lviv for Kyiv, to train the Ukrainian army there, one of the people I was working with received a call to say that the American government was looking for a team that could get an American citizen out of the warzone,' Neal tells me. 'That was all they said. We had to accept or refuse the mission based on that information. We had no idea what we'd have to do, where we'd have to go, but we said yes. If an American citizen needed to be got out, then we'd do it. There were three of us, the coordinator, a young guy and me. The young guy had some kind of connection with the coordinator. If I understand correctly he was the godfather of the younger man's child. The lad was a Ukrainian marine. We would be introduced to a US representative. That was to happen the next day; they'd even paid for us to stay in a hotel.'

A light-green landscape slides past behind Neal's bearded face. He's wearing the Ukrainian camouflage army hat, from which he's inseparable. It suits him. He's bald underneath, and when he takes it off he suddenly looks much older. He let his beard grow after the local troops told him he had 'an American chin', which is to say narrower than the broad jaws they have here. He's identified himself in every way with the people of this country. With their heavy chins and skulls, the inhabitants of these parts, like the mayor of Kyiv for example, are good at boxing and mixed martial arts; the strength of your skull and jaw determines your resilience. With a light head and a smaller chin you're more quickly knocked out.

'We had to go to the lobby of the hotel in Lviv, where we'd find the representative of the American state. It turned out to be a woman. Long blonde hair, pretty face, good figure. If I'd come upon her in a bar, I'd have offered her a drink. She invited us to her hotel room to discuss the details of the mission. What she told us was that an American couple had adopted a child. The process was underway before the war began. Everything had been arranged, the paperwork completed. The baby was an American citizen. They only needed to wait for it to be born.'

Neal takes a moment.

Childless couples in America adopted children from Russia and Ukraine. In practice the women were more like surrogate mothers, since the deal was often made before they even got pregnant. When relations between Russia and the US cooled after the Magnitsky Act was passed, Russia banned any more such arrangements. Putin didn't want the enemy to benefit from Russian fertility. But in Ukraine it continued.

'The day after the birth, the hospital was bombed. The baby was in the maternity ward and the father was with the mother in an adult ward, which received a direct hit. The baby survived but the father and mother were killed. The idea had originally been that the biological parents would travel to Warsaw to meet the American adoptive parents. But that couldn't happen now. There was no way to get the baby to Poland. The adoptive parents in America were at their wits' end and they'd approached a number of organizations to ask whether they could help to get the child out of the warzone. One group of individuals was prepared to fetch the baby but wanted payment of 100,000 dollars. Since there was no way the adoptive parents could pay that, they approached the American embassy in Warsaw, in desperation. The baby, a boy, was an American citizen. That had all been arranged. He already had a name, Samuel, meaning God has heard.

'The Americans eventually homed in on the coordinator, who looked for "the talent" that could get the baby to Lviv alive. The baby was a little more than 1,200 kilometres away, in a warzone, in a secret underground hospital in Mariupol. After everything had been discussed, we were made to sign an NDA, a non-disclosure agreement. Nothing about the mission was allowed to come out.'

Neal lights a cigarette. I don't mind him smoking in the car as long as he opens a window. He's reasonable and usually smokes only when we stop, but sometimes he can't hold back. I don't know whether it's a matter of tension or of routine. Everywhere in Ukraine people still smoke as if their lives depended on it. That will be a stumbling block when they join the EU. While he smokes, up against the window, his lips almost offroad, the story stalls.

Neal has finished his cigarette. He stubs it out and keeps the butt to throw away later. He's internalized the 'leave no traces' rule; in enemy territory it can make the difference between life and death.

'The coordinator came up with the idea of using an ambulance. That way we'd quickly be allowed through roadblocks, and with the siren turned on we'd be untroubled by congestion and traffic lights, able to drive smoothly through intersections and cover the 1,200 kilometres in as short a time as possible. The next day a Swiss friend of the coordinator gave us a choice between two Swiss ambulances. One was a big van and the other a smaller Volvo. We chose the Volvo. It was ideal. And fast.

'We put on paramedics' uniforms, green, with yellow reflective stripes on the trouser legs and shoulders. Underneath we wore bullet-proof vests. We swept everything out of the medicine cabinets and stuck our weapons and ammunition into them. All three of us had an AK-47 and a lot of extra ammunition. We left the gurney in the back. We took the shortest route through what was said to be safe territory. All the same, we had to spend several hours in areas that were mainly under Russian control.

'The biggest danger were the checkpoints. The Russians attacked them regularly, shot everyone dead and then

manned them wearing the green or blue armbands of the Ukrainians. The ambulance gave us the best chances there too. We drove at speed through towns and cities with the siren and flashing lights on, and had fun doing that; it was almost childish. After driving for twenty hours we arrived in Mariupol at about midnight. Our trip had gone without a hitch, no firefights, nothing.

'I was in the back, my weapon at first in the medicine locker above me and later under the gurney, so that I could get to it quickly. At checkpoints I was able to hide behind a cabinet. There was a minimal window in the back door. If I made myself small I could avoid being seen through the glass. At every checkpoint we drove straight to the front of the queue with our flashing lights and siren and were waved through immediately. In Mariupol we parked the ambulance in a suburb. Our worry there was that someone would give us away to the Russians, who would send a patrol after us. There are still people in Mariupol who sympathize with the Russians.'

In Western media you get the impression that all Ukrainians are united in their hatred of Russians. When I asked a few people about that in Kyiv, all of them very pro-Ukrainian, three told me that in the whole of Ukraine, including Luhansk, Donetsk and the Crimea, around ten per cent are pro-Russian, with the biggest concentration in the three occupied areas where, especially since 2014, the Russians have brought in a lot of people from Russia proper.

Lviv, Monday 30 May 2022

The baby from Mariupol

We walk through Lviv. Gathered around a bus stop on a broad pavement, middle-aged women and grannies are selling potatoes, onions and small tomatoes out of cardboard boxes. Cut flowers, arranged by colour, stand in plastic mayonnaise buckets. On a piece of plastic spread out on the ground are purple calyxes and petals from chrysanthemums of some kind. It's busy in the narrow streets of Lviv.

Neal and his two companions liberated baby Samuel from Mariupol on 3 April 2022. It was a mild night, nine degrees Celsius; two weeks earlier, the city was still experiencing five degrees of frost. I looked that up in order to be able to imagine the circumstances rather better. On a military map dated 1 April there were still two parts of Mariupol not yet in Russian hands, the Azovstal factory with a long adjoining stretch of territory from the sea to the north-east, and the Prymorskyi District. On the first of April the Russians attacked Mariupol from three directions. According to a map by southfront.org (a Russian site), Prymorskyi was attacked from the north and the Azovstal factory and its surroundings from Zhovtneve and Ordzhonikidzivsky. The next map, published by that same Russian site, shows the situation on 7 April. The area controlled by the Ukrainians has shrunk appreciably. Perhaps the maps give a favourable impression of Russian progress, but there's no room for doubt that in those days, and therefore on 3 April, there was heavy fighting in Mariupol. The Russian forces made significant advances and were present everywhere in and around the city.

The British ministry of defence, a well-informed source in this war, puts out daily intelligence updates. The map it published for 3 April shows that from Zaporizhzhia down, practically all of southern Ukraine had been taken by the Russians, with just a few irregularly shaped crosshatched areas where it was unclear which of the warring parties had the upper hand. The shortest route from Zaporizhzhia to Mariupol is 227 kilometres. I don't know exactly what route the ambulance took and neither does Neal. As far as possible they used minor roads and they had to drive for hours across territory where they didn't know whether the checkpoints were manned by Ukrainians or Russians, through areas in which at any moment they might come upon a Russian patrol or column.

'The other two were cool as cucumbers. In Mariupol we took turns keeping watch. Just short of Zaporizhzhia we stood the AK-47s right next to us, between the doors and the seats. I drove the first stretch, to well past Kyiv, but as soon as we needed to pass Ukrainian and possibly even Russian checkpoints, the coordinator and the young marine took over. They could speak both Russian and Ukrainian. I took a seat at the back, and through a small window between the driver and the patient compartment I could communicate with them, and shoot if necessary.'

In Lviv there's still asparagus on the menu. It's available and I order some. It surprises me how trade continues as normal in wartime. The asparagus was grown near Kherson, a good agricultural area.

'At first light we started driving, from the suburb to the place where the underground hospital was supposed to be. A large U-shaped brick building had been bombed and had collapsed. Only a few fragments of wall were still standing; other than that it was a big heap of rubble. Hidden among the ruins, a flight of steps led down to two metal cellar doors that gave access to an improvised underground hospital. It was dark inside. There were puddles of water. In the beds were hundreds of wounded.'

I ask him whether they took the Kalashnikovs into the hospital with them.

'What do you think? We were in Mariupol!'

He looks at me in amazement. We're sitting in a courtyard, beneath grapevines. I vividly imagine the three men dressed as paramedics, in green uniforms with reflective stripes on the trouser legs and Kalashnikovs over their shoulders. They look rather stout because of the bulletproof vests under their jackets.

'People were expecting us. Almost immediately a nurse appeared out of the darkness with a baby in her arms. She gave us the baby. And a travel cot, a flask of warm water and some milk powder. That was good, because we hadn't thought of any of that; we'd mainly stocked up on lots of ammunition and water, and a little food. The coordinator took the baby into his arms. It was wrapped in blankets.'

How on earth do you hold both a six-day-old baby and a Kalashnikov? Neal explains to me that it's not such a problem, if you hold the weapon in the most convenient way, with the barrel pointing downwards. You can fix the shoulder strap of a Kalashnikov to either one point or two. He prefers one point, since you can react more quickly and aim in all directions. From the moment he tells me that, I pay more attention and see that most Ukrainian soldiers opt for Neal's method, fixing the shoulder strap to a single point on the gun. It gives them the maximum freedom of movement and it looks cool, letting your AK-47 dangle low. The Ukrainian troops look unbeatable. They radiate indomitability, whereas many Russian troops go about in mismatched clusters.

'It was a smooth trip there and a smooth trip back. We took the baby and drove the whole way to Lviv in one go. No problems, no Russians, no firefights. We tied the travel cot to the gurney. In the hospital they'd given us some nappies.'

Three battle-hardened men with AK-47s, a baby and baby milk in an ambulance. I ask how they managed to change nappies. Who did that?

'We took turns. The baby slept for most of the journey. He behaved impeccably. I think because of the rocking and swaying of the ambulance.'

The Ukrainian roads were not universally smooth before the war, to put it mildly. They certainly haven't improved since. Heavy transports and tanks ruin the road surface, the asphalt is torn off by machine-gun fire and there are craters made by incoming mortars and rockets. Maintenance has to wait for the time being.

'When the baby cried he was hungry or needed his nappy changing. We gave him the bottle and then a tap between the shoulders to make him burp, and he'd go back to sleep. Samuel was absolutely a model baby.'

Neal beams from ear to ear. The sight of him and how he tells this story breaks my heart. Here is a man who has killed I don't know how many people in his life and now he had been on a mission to save someone.

A six-day-old baby without parents in Mariupol under Russian siege – that's not a favourable prospect. Neal is the father of three children and I see from his gestures that he's cared for babies and toddlers. I recognize this. I made up for the lack of a father with my own three sons, doing as much as possible with them. Whenever I got frustrated because it was impossible to get around to writing, Ilona would say, 'I don't think you'll ever regret the time you've spent with your children.' And she was right about that. I'm a Twentenaar, a man of the clan.

Neal shows me how he held the baby to his chest, giving it an imaginary slap on the back to release a burp. We eat our asparagus.

'It's surely incredible that you took the decision to go and fight in Ukraine when you saw a maternity hospital in Mariupol being bombed and then saved a baby from an underground hospital there of all places,' I say. 'It gives a beautiful shape to your mission.'

Neal puts down his knife and fork and looks at me.

'The war here has had a healing effect on me. It's washed me clean and cured the wounds of Iraq.' He nods for a long time, thoughtfully. 'Iraq was horrific. We didn't have a minute's peace. The men there couldn't wait to die and take their children and wives with them into death. Car bombs packed with children drove at us. You had to open fire on cars with whole families in them. I joined the Marines at eighteen because I wanted to fight evil. I've been able to do that here. I'm grateful to the Ukrainians, deeply grateful. They're tremendous. I'm so glad I came here and was able to help. I'm leaving a whole lot of ballast behind in Ukraine.'

We raise our glasses in silence and drink. I ask whether he's met Samuel's adoptive parents.

'We drove all the way from Mariupol to Lviv and met them in a hotel. In the same hotel, in the US official's room. They'd come from Warsaw. The father was extremely grateful to us, the mother hysterical. When we walked into the room with the baby she stood up, put her hands in front of her eyes, and tears poured down her face. She simply couldn't stop crying. She stuck out both her arms. I don't believe she could see anything yet. The blonde US official took Samuel out of our arms and laid him in the arms of the adoptive mother.'

Medyka (Poland), Wednesday afternoon, 1 July 2022

Three bags of ladies clothes

'What an honour. Just to be in that room along with all those brains!'

The great thing about Neal is his sincere enthusiasm. This morning, on our way westwards from Lviv, we visited a group of drone builders. They are engineers, drone pilots and brilliant hobbyists, all brought together under the leadership of a former investment banker. They build drones and robots in an old factory. The robots are small tracked vehicles that they've developed using simple materials. The caterpillar tracks are made of rolls of plastic with rows of blind rivets for grip. The robots can take medicines to a wounded serviceman or to a doctor on the battlefield, or food to the starving, or – and they're working on this – clear mines by sweeping long arms. Neal has explained to me that the rule of thumb for mine clearance is that you have to multiply the number of days a war has been fought by three to find the average amount of time needed to make the country free of mines. I and a group of friends have supported the drone-building whizz kids financially. They order parts online from all over the world and between the bare walls of an old brickworks they put together one reconnaissance drone after another. Drones are playing a crucial role in this war.

Through truly delightful rolling hills we drive towards the Ukrainian-Polish border. We're in the pickup, which has a winch to haul freshly killed deer and wild swine into the back. As far as documents are concerned, five nationalities are associated with this vehicle. The license plate is Hungarian, the stickers on the windows and the stamped papers from Kyiv say that I work for a Ukrainian foundation domiciled in Poland, I'm travelling on a Dutch passport, and my passenger has an American passport. The biggest potential stumbling block at the border, I reckon, is that Neal has fought as a sniper for the Ukrainian Foreign Legion. Hungary does not allow the transport of weapons into Ukraine from its territory. So it seems to me that the best way to enter Europe with an American sniper is via Poland. I don't feel like being needlessly detained for twenty hours at the Hungarian border. I don't know exactly what Neal's got in that heavy rucksack. The Polish border guards, with their shared aversion for the Russians, will be more favourably disposed. I can then drive without any difficulty into Hungary via Slovakia.

'It's a militocracy, a warrior people. They respect force. They respect it when you know how to handle weapons,' Neal says, and he tells me how after he arrived at the training centre in Kyiv he had to fight a giant, a Ukrainian karate champion. A big circle of troops stood around them. The giant was quick as well as big. Neal got the full weight of a fist on his chest, which bruised his ribs. He was racked with pain, but with an open hand he made a move towards the giant's eyes and at the same time swung his leg to kick him as hard as he could on the knee. For two weeks he barely slept because of the bruised ribs and the pain in his foot. The giant was unable to walk for several days. Neal had won the respect of the karate champion and of the men. He and the giant both smiled appreciatively whenever they ran into each other, limping.

I look around. Since I came to live in Eastern Europe, almost twenty years ago now, history has become much more vivid to me. Because history makes it easier for me to understand something about the people I live among and over whose land I'm travelling, but also because history continually comes into the foreground, far more so than in the West. In the Netherlands we've developed a sanctimonious attitude, a lie as I see it, which says that history doesn't matter. You are who you are, as an individual. It's seen as insulting to ask a person about their history, about where they come from. To me those are essential questions: where are you from, who are you, where are you going? Without knowing the answers, you can only circle around each other with superficial insouciance. How can you understand another person without knowing those things?

We're approaching the border. I must quickly leave behind me the enjoyable lawlessness, the abandonment of all refinements – including speed limits, parking bans and respect for solid white lines on the asphalt – and start obeying the law again. The road to the border is fairly empty, but near the crossing at Przemyśl it's full of buses, vans and ambulances. There's an amiable atmosphere, with transport in both directions; most drivers seem to know the ritual. Neal points out to me the white tent of the Foreign Legion to which he reported, months ago.

On the Ukrainian side of the border all kinds of papers have to be checked and stamped, all in a specified order. You go from one waist-high hatch to the next, needing to give a deep bow at each one, a relic of the old Soviet system of humiliation. The administrative rigmarole is old-fashioned Eastern Bloc and we advance laboriously, in a caravan of vans and buses. There are not many cars. Most vehicles in both directions are trucks, although not army trucks. My all-terrain vehicle with mud tyres is a bird of paradise in the queue. A considerable number of border guards are walking around with Kalashnikovs. Neal remarks drily that none of the Kalashnikovs has a magazine in it, nor are the guards wearing magazine pouches.

The Ukrainian border guards order us to drag our bags out of the pickup and put them on a long, wide wooden bench. The people in the vehicles ahead of us have to do the same. I've already put a lot of our luggage into the bed of the truck in the hope that it could be checked there, but that kite doesn't fly. Everything has to go onto the low bench, so that the entire queue behind us can see what's fished out of our bags. From Neal's fat green rucksack come a helmet and a bullet-proof vest. Out of my bag come underpants, laptops, chargers, maps, notebooks and an exploded Russian shell. I found it on the road to Chernihiv. Neal decided it must have come from a blown-out armoured personnel carrier.

A huge uproar arises over the shell and eventually, like a schoolboy, accompanied by a soldier with a Kalashnikov lacking a magazine, I have to throw the rusty exploded shell into a garbage bin. It's clear that the Ukrainian side has been tasked with ensuring that no weapons or bits of war memorabilia get into the European Union, since the Polish guards look only at our documents, barely glancing at our luggage. In Hostomel and Irpin green ammunition boxes had been left lying along the hard shoulder and at the roadside. I didn't stop, and that bothered me for days; I'd have loved to take one or two of those boxes as a souvenir. But I'd probably have had to leave it here.

The customs officers become suspicious as soon as they start digging through the three bags on the back seat. A friend of Margarita's went to her flat in Kyiv, filled three bags with clothes, carefully wound tape around them and delivered them to me at Hotel Hermitage. I haven't opened them. I've taken it on trust that they contain clothes. They're now sitting on the special wooden inspection bench. They're those huge rectangular bags with a zip at the top and big handles.

The border guards remove the heavy tape. While they search I stand right in front of them to make sure nothing gets filched. I watch as white trousers, floral summer dresses, silk blouses, lacy lingerie, transparent underwear, Dolce & Gabbana bras, negligées, bustiers, suspenders and wafer-thin pyjamas are passed through the callused hands of the customs officers. Padded boxes containing watches, earrings and pearl necklaces are opened and then very carefully closed again, the customs officers holding their breath as if they've become privy to a secret. I too have the sense that I'm catching a glimpse of something not intended for my eyes.

I don't know quite where the problem lies, but the combination of Margarita's clothes and the American sniper I have with me is not ideal from a customs and border protection perspective. Two unshaven, stinking men returning from a warzone in a dusty all-terrain vehicle with bags full of gossamer-thin ladies clothing, a rucksack containing a kevlar helmet and a bullet-proof vest, eight empty jerrycans, two Kyiv cakes and an exploded shell. It's incongruous.

After all the bags have been inspected we're allowed to repack them, and the officer on duty is called over. He takes us to a shed made of corrugated iron and starts speaking Russian. After a while it dawns on him that Neal doesn't speak Russian and he switches to rudimentary English. To the question of what he's been doing in Ukraine, Neal lies with a pious face that he's been working as a nurse, forgetting that the 'master hunter' medal is still proudly displayed on his chest, a rare Ukrainian honour reserved for snipers. After I've convinced him that I too don't speak Russian, the officer gestures that I must park the vehicle off to one side. We can wait by the car. He takes our passports with him.

Fortunately, online background checks on great hulks of discoloured computers produce nothing alarming and after four and a half hours we're let through. We drive into Poland and forty-five minutes later we're in Slovakia. We cross hills and forests. In broad daylight I see a roe deer bounding through tall grass in the direction of the road. I warn Neal, who brakes, and the deer crosses in front of us unharmed. He nods and says, 'There were a lot of those little deer in Zaporizhzhia as well.'

As we drive back through Slovakia to Budapest, I feel gloomy about the war, especially about the meat-grinder tactics of the Russian army in the east. They simply bomb everything to bits. That way the Russian soldiers don't need to fight – which they're not eager to do. Aside from a few professional raiders like the men of the Wagner group, they have no motivation. I'm assailed by doubt. Is there any point in all the efforts I'm making to get protective equipment to the army amid this devastating violence? Am I fooling myself?

The next morning, at home in Budapest, I receive an encouraging message from Dima, the man who sent me the list of supplies needed at the very start of the invasion:

> *Good morning, Jaap! Just want to let you know that one of the helmets you guys funded saved the life of one of my friends a couple of days ago. He's on the front line near Izyum. A fragment of an artillery shell hit the back of his head. He'd be dead had he not had a helmet.*
>
> *Thank you for doing this,*
> *Dima*

Epilogue

Lónya, August 2022

Mother Ukraine

'You understand that I can't take any responsibility at all for Ravenna,' I tell Zita, my wife's younger sister.

'We realize that,' Zita says softly. She's also speaking on behalf of her husband. 'We know it's a country at war. It's Ravenna's own responsibility. She wants this very badly.'

'I'll try not to do anything too stupid.'

It's quite something, taking your twenty-eight-year-old niece with you into a war-torn country. The likelihood of being hit by a cruise missile in Lviv or Kyiv is not great, true, but my main worry concerns a small-scale nuclear attack or the Russian shelling of a nuclear power station. I say nothing about that for now, to avoid disquieting Zita's maternal heart unnecessarily, in the hope that she won't toss and turn in her sleep for ten nights. Everything I see and hear from my Ukrainian contacts tells me that the Russians can never win this war. Only scorched earth tactics, the tried and tested Russian method, can offer them any consolation. Using a tactical nuclear weapon or triggering a nuclear disaster at Zaporizhzhia to create maximum chaos in Ukraine and in Europe would be a logical part of that.

Along with Ravenna, Zita's oldest daughter, I'm going into Ukraine to help make a documentary called *The Art of Survival*, about the role of art in war. In our village in Somogy we discuss the plan in the evenings, do some research and prepare for our trip.

In early August, Ravenna and I drive to Kyiv via Uzhhorod and Lviv. At the airport in Budapest we pick up Alex, a highly experienced sound technician, born and raised in the Crimea. Having grown up speaking Russian, he's now learning Ukrainian. He's been in the Ukrainian special forces and has lived for the past quarter-century in Amsterdam.

Along the way and in Kyiv I'm struck by the fact that the sandbags used to create walls and fortresses are falling apart, the white sand trickling out of them. The plastic is torn. It's unravelling, the way everything is starting to unravel to some extent – a metaphor for what's happening on a larger scale, in Ukraine, in the world. The war is six months old. Coalitions are beginning to show cracks, alliances falling apart. Enthusiasm is waning. Interest is ebbing away. It's getting tiresome, inflation is rising, energy prices exploding. This is unsustainable; people have had enough. The shrewd Russian propaganda – 'this is not our war' – is catching on in Europe with the extreme right and with those having a tough time because of high energy prices.

Even for people in Ukraine who stand foursquare behind the army it's impossible to keep going. There's no money left. People tell me they're eating less because they can't afford food. Putin knows all this perfectly well. He hasn't managed to win the war on the battlefield, so now he's trying to win by causing maximum chaos all over the globe. Illya explained to me, when we saw each other a few weeks ago, the difference between the Mafia and the Russian secret service: 'Criminals ultimately have certain rules; the FSB doesn't. All that counts is the mission. In fulfilling your mission everything, absolutely everything, is permitted.'

There's an app that warns people of bombs and missiles in Ukraine. In real time, as it's called, you can see on the screen of your phone where there's a danger that a rocket might hit. The provinces that might be on the receiving end of cruise missiles coming from Russia or the Crimea are coloured red. As soon as there are impacts, the app shows where explosions have occurred. You can adjust the volume of the air-raid siren. We installed the app on our phones around three quarters of an hour out of Kyiv. It got dark and the alarm went off. On the screen the whole of Ukraine, aside from Transcarpathia, turned red. I felt some slight fear in the car. While we were in Kyiv, the alarm sounded an average of five times a day. Offices had to be emptied and shops, cafés and restaurants shut. Suddenly the streets became pleasantly busy. Hardly anyone goes into the bomb shelters, since you'd be up and down there all the time if you did.

Back in April, Neal wrote to me that what happened in Bucha and Irpin, the dragging of civilians out of their houses and torturing and dishonouring them at specially equipped places, is standard, a method applied by the Russians everywhere.
Ravenna and I go to Borodianka, up towards the border with Belarus. While the Ukrainian cameraman films the destruction, I go into an abandoned block of flats and pick my way through devastated apartments. The residents have fled. Some were killed. The atrocities committed in Borodianka are comparable to those of Bucha. I put one foot carefully in front of the other. In the stairwell a beautiful big old Bible is leaning against the wall, left behind by someone because it was too heavy. Or left behind by the Russians, booby trapped, as I realized later. The intimacy of the homes where the front doors have been blown off is enchanting and terrible. Others have been here before me. Every cupboard, every drawer has been opened, every electronic device has been taken. The floors are littered with packaging, curtains, net curtains, clothes, medicine boxes and slashed cushions. It's extremely depressing to shuffle through all this destruction. Everything has been turned upside-down. Only the bookcases have been left undisturbed.

Outside Borodianka, trenches have been dug and underground bunkers built by the Ukrainian army. There's a fear of a fresh attack on Kyiv from Belarus. Along the crash barrier are dozens of signs warning of new minefields, laid by the Ukrainians. On the way back to Kyiv we pass a forest of scrawny spruce trees with a burned-out column of Russian tanks in it. I scour about in the woodland next to the tanks and come upon the place where the Russians made camp. A pigsty, as it is everywhere the Russians have been: the burned sole of a heavy boot, broken vodka bottles, throwaway razor blades, toothpaste, toothbrushes, underpants, food scraps, packaging. Between the camp and the tank I spot a human upper jaw with six teeth – one of the front teeth is missing – and a fragment of lower jaw.

On 28 February I wrote that it was perhaps a hopeful sign that Sergey Naryshkin, head of the foreign department of the Russian secret service, was not exactly howling with the wolves in the forest that it was a good idea to invade Ukraine. At that point he was almost certainly the best-informed man in the world on the state of the Russian and Ukrainian armed forces. He knew how corrupt the whole army was, from the middle ranks to the top; he knew about the beautiful second homes the Russian generals have in the Crimea, palaces their salaries could never have paid for; he knew that the troops were wearing summer uniforms and even had to buy their own bullet-proof vests, and that they were drinking the fuel for their tanks to keep warm, in lieu of alcohol. With its major losses at Izyum, Russia has become Ukraine's biggest arms supplier, certainly when it comes to offensive weapons.

After a week spent filming, Alex and I head back towards Hungary via Lviv and Uzhhorod. At that point *ZOV* appears on the internet, a 140-page report by a Russian paratrooper called Pavel Filatyev. He took part in the Russian occupation of Kherson and fled abroad after a few months. He describes in detail the total chaos and corruption of the Russian army. He had to buy his own boots, and after waiting for ten days he was given a summer uniform in winter. He describes the apathy among the soldiers and the way ninety per cent of the troops talked in the smoking room about how to leave the army as quickly as possible – even though they were all part of an elite unit. According to Pavel, the Russian army is a madhouse where everything is for show. He writes that almost everyone is suffering from fungal infections to their feet. When they arrived in Kherson he was so hungry that, like the others, he pulled open kitchen cabinets in random houses and, like a looter, tipped out the contents to find something edible.

The athlete's foot infections remind me of Neal, of course. Weeks ago he alerted me to the importance of good hygiene, explaining why he took baby wipes and spare socks with him on his sniping missions. He often said that the Russian army would fall apart in the winter, because it's so chaotic and the middle echelons are enriching themselves by stripping the supply warehouses, so the troops can't get hold of the materiel they need to survive. 'In the winter the Russians will die like flies,' he said.

He's back in Alabama. There he was received as a hero and he's now mowing grass again and tending gardens in the suburbs of Crane Hill. In Alabama he can speak freely. The day after the rescue of the baby from Mariupol, the blonde representative of the American government wrote about it at length on social media. Neal asked her whether that meant the non-disclosure agreement had ended. Indeed it had. He could talk about it as long as he didn't give the names of the adoptive couple. He was even allowed to divulge the baby's name, Samuel. He's busy. He sent a cryptic answer to a WhatsApp message from me a few days ago.

> *I'm doing fine. Thank you, brother. My Ukrainian family is doing well. A few wounds from Russian rocket attacks, but aside from that the family is still complete. I've asked them whether they can help me to buy a house or an apartment. When I can, I'll move there. I love Mother Ukraine and I love the people. I've got a family there and they're calling to me to come home.*

The message throws me a little. Does he literally have a family there? Did he fall in love with a Ukrainian woman during his months in Zaporizhzhia, Kyiv or Mariupol, or by 'family' does he mean the soldiers, the brotherhood? On our road trip to Kozelets, Chernihiv, Lviv and Budapest he was cheerful and deeply grateful to the Ukrainians. I've sent him a message in response, asking what he means by his family in Ukraine.

The border crossing at Záhony is a disaster. We line up to get out of the country on a kind of square in front of the Ukrainian border post in the sweltering heat and barely make any progress. All the cars have their doors open to let the breeze blow through. Roma children go from car to car and stand there until they're either given something or sworn at. For the past half hour we haven't moved an inch. With Alex I walk over to the first guardhouse. To my amazement we're not barked at to go away; the customs officer tells us it's dreadfully slow on the Hungarian side of the border. He doesn't know why. There are six cars still in front of us, but you know for sure that several hundred metres away, on the other side of the steel bridge I drove over two months ago, there are many times that. After an hour and a half I've had enough.

'Doors shut, we're off,' I say, and I turn the car out of its lane. On the square I perform a U-turn, mounting the pavement with the heavy mud-terrain tyres. We drive towards the little border crossing at Lónya, which I know from my previous visit to Transcarpathia. There wasn't a single person there that time. It's the forgotten border crossing where after the border post you can take the little ferry across the Tisza. I've used that ferry three times before; it costs 500 forint with the car, or 1.25 euro, and you're dragged across the brown water of the river along a thick steel cable. Last time, this border crossing took me fifteen minutes. Ravenna and Alex have their doubts, to say the least, about the wisdom of my impulsive act.

We drive through empty villages with beautiful churches and fruit trees at the roadside. Not a soul to be seen. It's utterly rural here: a few vineyards, lots of maize, bare soil, brambles beside the road. A long, narrow strip of asphalt, a road to nowhere, leads to the border. An old Lada comes towards us with a rectangular sign on the roof, a hire car. We pass a tall watchtower, painted light blue, and a little later we reach the Ukrainian border post, a sluice of concrete blocks with eight or so customs officers in uniform and soldiers in camouflage, including two women with Kalashnikovs. There's no queue, there are no other cars, only a man on a bicycle with his walking stick strapped to the rack. Peace has returned to the border post at Lónya. We're soon past the Ukrainian border.

After checking the papers for the Toyota, the customs officers tell us we'll have to turn round. The vehicle we're in has the status of a truck, it seems, and since the start of the month no trucks have been allowed through the border here. That's to say, we can pass the Ukrainian border but not the Hungarian. The law has changed in Hungary. With my impossible optimism I say we'll give it a go. Alex and Ravenna frown. The Ukrainian border guards shrug.

We drive across no man's land between the two border posts. It's about a kilometre. I have a sense we're as good as through and feel triumphant; leaving the queue in Záhony was the right decision. The branches hang low over the asphalt. The shrubbery next to the road is impossible, thick and full of brambles and other hostile flora. Tasya, the cat belonging to the children of Alexey and Yeva, escaped from the VW van here and disappeared into the jungle. Yeva had to crawl after it on hands and knees and beg the cat to come back.

Ahead of us the Hungarian border post looms up. There are five cars. That's more than I'd hoped. This is where Alexey and Yeva stood six months ago; this is where my whole Ukrainian adventure began. It's unreal to be at the Hungarian border post at Lónya and see it so empty, in the scorching heat. I vividly recall the endless throng of people standing here in the freezing cold. All gone.

Leaving the country makes me melancholy. I've become attached to the people I've met in the pressure cooker of the war.

The airport at Odesa was hit by rocket fire from the Crimea on 1 May. It's the airport where Alexey was installing an air traffic control system the day before the war started. The next day the system was due to be completed, after months of effort. I suspect that Alexey isn't too bothered about seeing his work destroyed. He's now making pizzas and hamburgers in Cannes, working his socks off in the Lucky You Saloon at the harbour. He's applying the analytical skills of a network engineer to the use of ovens and to fathoming out ideal cooking times and temperatures. Yeva gets photography assignments on the Côte d'Azur and earns as much from each one of them as Alexey gets for a month working at hot ovens. Alexey says he wants to return to Kyiv with Yeva and the boys as soon as peace comes, that he's longing to start using his brain again.

My niece Madelien, who shares her life with Anton, Alexey's younger brother, had a baby in June, a girl, Valentina, so the Dutch-Ukrainian family is growing.

Margarita is continuing to work for the same Ukrainian firm from Budapest for the time being. She dreams of a family and occasionally rings her mother. I handed the three huge bags of clothes to her in Budapest two months ago.

'I don't talk with my mother about the war, we avoid that subject, because it makes her aggressive. And me too,' Margarita writes. 'She's gone back to Pervomaisk in Donbas, I've no idea why. We still have an apartment there, and a house. There's a bit of shooting and from time to time it's bombed; most of the neighbours have left. We talk perhaps once a week, about the weather, about the cat. I can't talk to her about anything important. To be honest, I don't like calling her, I find her repulsive, but she's still my mother and I have to respect her. I'm waiting for an offer from a company in Hamburg. I love travelling, I love Europe; I've always dreamed of travelling and of being in Europe and now here I am. You see, a person needs to be careful what they wish for.'

Yulia, the director of the cake factory, spends her time with the commander and is still busy running things. She delivers Kyiv cakes and love to all the furthest flung parts of the motherland, to children at school, soldiers at the front and the wounded in the hospitals, from Kharkiv to Mykolaiv, and tears come to her eyes whenever she sees a baby. At the front close to Mykolaiv she was bitten by a steppe viper, but fortunately she received treatment in good time.

Illya is in Poland, making music again, teaching at a music school, and in his spare time helping volunteers to deliver things to the front. He tells me it's becoming increasingly difficult to get hold of supplies, to find trustworthy parties. There are more and more people trying to earn money from the war. He also visits badly wounded soldiers in Polish hospitals now and then. The men who have lost arms and legs cheer him up with their jokes.

Three days before the invasion began, Vlad delivered the manuscript of his new book. *The Pyramid Mind* will be published by American publisher Simon & Schuster. I'm in contact with him from time to time. He's on a secret mission. We tried to meet up in Kyiv but it didn't work out. He's somewhere else. He can't say where.

At Dirck's farm the winter and summer wheat and the rapeseed have all been harvested. The buckwheat needs to be got out of the country quickly. Two of the men from his village have already been killed at the front. Of his fifteen employees, not one has been called up as yet. They are essential to bring in the harvest. It's impossible to book rail freight wagons and road transport has become unaffordable. The sheds and silos are full, and prices are low for agricultural products from Ukraine.

I congratulate Anastasia, the singer, with the successes of the Ukrainian army around Kharkiv. She writes back, 'Yes, but at what price?' Anastasia and Sasha no longer live in the music studio. It's being rebuilt. All the musical instruments and recording equipment have been removed. Anastasia travels back and forth to Poland and is trying to ar-

range for Sasha to come with her. They're still fully devoted to voluntary work. Sasha sometimes performs with his band. His drummer is at the front. They tell me that a lot of their friends already have streets in Lviv named after them – that's what it's called when you die at the front. In Lviv young men are commemorated daily. I've been there with Ravenna. A large crowd gathers in front of the town hall. When the car with the coffin arrives, the whole square falls deathly silent. The priest gets out. A trumpeter in front of the town hall starts to play. Behind the hearse is a bus carrying family members. As the procession sets off again, everyone on the square bows their head and makes the sign of the cross.

Phaeton is refurbishing machine guns in Zaporizhzhia and from time to time has to go to the front, where he wears a level four bulletproof vest. He asks me whether I'll come to celebrate victory when it arrives. I certainly will. From time to time he sends me a photo of a machine gun that he's restored, the way someone else might send me a post card showing a mediaeval church.

Ella, the barista from Kharkiv, didn't stay in Turkey. She's now living and working in a hotel in Wexford, Ireland. The girls she brought together four months ago are scattered across the world. Olha is now in the United States. Valeriya with the blonde curls and Maryna, the youngest of the three who didn't say a word, spent time in Lithuania and in Turkey and are now together in Vienna. Valeriya is working and Maryna is studying online at a Ukrainian university.

Ella hopes to return to Ukraine soon, but there's heavy fighting in her city. She tells me that men in the Kharkiv region were forced to undress in front of Russian soldiers so they could be examined for nationalistic tattoos. She's in touch with friends who stayed in the city, but she tries not to follow the news too much because it's depressing. She doesn't mention how her son is doing.

This morning I got a reply from Neal:

My soldiers are my family in Zaporizhzhia. My commander from Kyiv was transferred to the front and killed by tank fire. That causes me intense pain; he was a fantastic person. It breaks my heart. He leaves a delightful wife, son and daughter. Of everything I've experienced, the loss of him affects me the most, to the depths of my soul. He was a good man. He didn't deserve to die. His family don't deserve to have to live without him. When the showers didn't work in our building, he drove me to his house so I could use his. He kept inviting me to his office and asking me how the training was going. We drank together. He told me the war was a heavy burden on his family. But he was passionate. He was fighting for freedom. When I got the message that he'd been killed... I'm a hardened man, Jaap. I can deal with anything... When I heard he'd been killed I cried like a baby, for hours. I'm crying as I write this. I will miss him. I do miss him. I'm going to bed now, my friend, sorry.. Speak to you soon.

We wait in line for an hour and a half. It takes about twenty minutes for each car to cross the border. When it's finally our turn, the Hungarian border guard says we'll have to go back. I tell him I came through here a few weeks ago, that it was no problem at all. No, the law has changed and there are cameras all over the place. He can't possibly let us through. I've gambled and lost.

In deathly silence we drive back along the overgrown road. It's unreal, like a film when there's suddenly no sound. My ears are ringing. I'll never get out of Ukraine; I'll never get done with Ukraine. Neal would be envious of me. In the thorny no man's land we float between two worlds. The Ukrainian border guards start to smile when they see the big green pickup approaching. The young soldier who gives us a raggedy piece of paper that has to be stamped by the various people responsible if we're to get through to the other side shortly can't suppress a broad smile.

'Ah, back again are you?! What a surprise! Can't you people live without Mother Ukraine?'

JAAP SCHOLTEN is a Dutchman who has lived mainly in Hungary since 2003. He is married to the daughter of a Hungarian refugee from the 1956 Revolution, and has travelled Central and Eastern Europe extensively. He is an award-winning author of travel stories and novels including *Tachtig, De wet van Spengler* and *Suikerbastaard*. For *Comrade Baron* he was nominated for the VPRO Bob den Uyl Prize and awarded the Libris History Prize. His books and stories have been translated into English, German, French, Hungarian, Croatian and Romanian.

"This is a classic in the lines of Patrick Leigh Fermor"
　　　　　Norman Stone, professor of modern history, Oxford

"I have enjoyed this book so much - such a great tale, with brilliant original research and source material, and so many stories, tragic, humiliating, painful, yet all engrossing and highly readable."
　　　　　Petroc Trelawny, BBC presenter and journalist

www.ingramcontent.com/pod-product-compliance
Lightning Source LLC
Chambersburg PA
CBHW030143100526
44592CB00011B/1018